The Roman Calendar from Numa to Constantine

The Roman Calendar from Numa to Constantine
Time, History and the *Fasti*

Jörg Rüpke

English translation by David M. B. Richardson. Originally published in German under the title *Kalender und Öffentlichkeit: Die Geschichte der Repräsentation und religiösen Qualifikation von Zeit in Rom*, by Jörg Rüpke, © Walter de Gruyter GmbH & Co. KG (1995).

WILEY-BLACKWELL
A John Wiley & Sons, Ltd., Publication

This edition first published in English 2011

The translation of this work was funded by Geisteswissenschaften International – Translation Funding for Humanities and Social Sciences from Germany, a joint initiative of the Fritz Thyssen Foundation, the German Federal Foreign Office, the collecting society VG WORT and the Börsenverein des Deutschen Buchhandels (German Publishers & Booksellers Association).

Edition history: © Walter de Gruyter GmbH & Co. KG (1e in German, 1995); Blackwell Publishers Ltd (1e in English, 2011 hardback)

Wiley-Blackwell is an imprint of John Wiley & Sons, formed by the merger of Wiley's global Scientific, Technical and Medical business with Blackwell Publishing.

Registered Office
John Wiley & Sons Ltd, The Atrium, Southern Gate, Chichester, West Sussex, PO19 8SQ, United Kingdom

Editorial Offices
350 Main Street, Malden, MA 02148-5020, USA
9600 Garsington Road, Oxford, OX4 2DQ, UK
The Atrium, Southern Gate, Chichester, West Sussex, PO19 8SQ, UK

For details of our global editorial offices, for customer services, and for information about how to apply for permission to reuse the copyright material in this book please see our website at www.wiley.com/wiley-blackwell.

Library of Congress Cataloging-in-Publication Data
Rüpke, Jörg.
[Kalender und Öffentlichkeit. English]
The Roman calendar from Numa to Constantine : time, history, and the fasti / by Jörg Rüpke ; English translation by David M. B. Richardson.
p. cm.
"Originally published in German under the title Kalender und Ölfentlichkeit : die Geschichte der Repräsentation und religiösen Qualifikation von Zeit in Rom ... Walter de Gruyter ... 1995."
Includes bibliographical references and index.
ISBN 978-0-470-65508-5 (hardcover : alk. paper) 1. Calendar, Roman. 2. Chronology, Roman. 3. Festivals–Rome–History. 4. Rome–Religion. 5. Rome–Social life and customs. I. Title.
CE46.R86 2011
529'.322–dc22

2010042237

A catalogue record for this book is available from the British Library.
This book is published in the following electronic formats: ePDF [ISBN 9781444396515]; Wiley Online Library [ISBN 9781444396539]; ePub [ISBN 9781444396522]

Set in 11/13pt Dante by Toppan Best-set Premedia Limited

1 2011

Contents

Preface

Interest in calendars, which grew markedly in the final decade of the second millennium of the Christian Era, has not waned in the third. Instead, we have seen an ever-stronger tendency to historicize calendars, and not only systems for reckoning time. The calendar has been revealed as both a medium of cultural memory and an arena for political debate. I am, therefore, delighted that my study of the Roman calendar and its graphic forms, published in 1995, can now appear in English translation. As regards the first part of that book, which introduced and discussed all surviving specimens of the Roman *fasti*, reference may still be made to the German version, which continues to represent the state of academic thinking. But the history of the Roman calendar, the second part of the earlier study, can now be read in new, revised, and more concise form, taking account of research carried out during the last fifteen years, both by myself and by others. I should like here to offer my special thanks to Denis Feeney (Princeton), Michel Humm (Strasbourg), and Benedikt Kranemann (Erfurt), for intensive discussions undertaken and suggestions offered, and also Clifford Ando (Chicago), Alessandro Barchiesi (Siena/Stanford), Jonathan Ben-Dov (Haifa), Hubert Cancik (Berlin), Gregor Kratz (Göttingen), John Scheid (Paris), Hermann Spieckermann (Göttingen), Sacha Stern (London), Olga Tellegen (Tilburg), Katharina Waldner (Erfurt), and Greg Woolf (St Andrews). The conditions offered by the research group *Religiöse Individualisierung in historischer Perspektive* gave me the opportunity to devote my entire attention to finalizing the revision process; for this, my heartfelt thanks are due to the *Deutsche Forschungsgemeinschaft* and the University of Erfurt, and the staff of the Max Weber Centre, in particular Bettina Hollstein and the director Hans Joas.

I am also grateful to Wiley-Blackwell for their readiness to publish this new edition. Haze Humbert made the way clear for this endeavour. David Richardson, to whom I already owe the meticulous translation of an earlier book, has repeated that service on this occasion, and helped to clarify the argument anew. Julia Carls, Alexandra Dalek, and especially Diana Püschel, helped revise the manuscript at Erfurt. Abbreviations used throughout follow *Thesaurus Linguae Latinae* for Latin texts and Liddell–Scott–Jones for Greek texts.

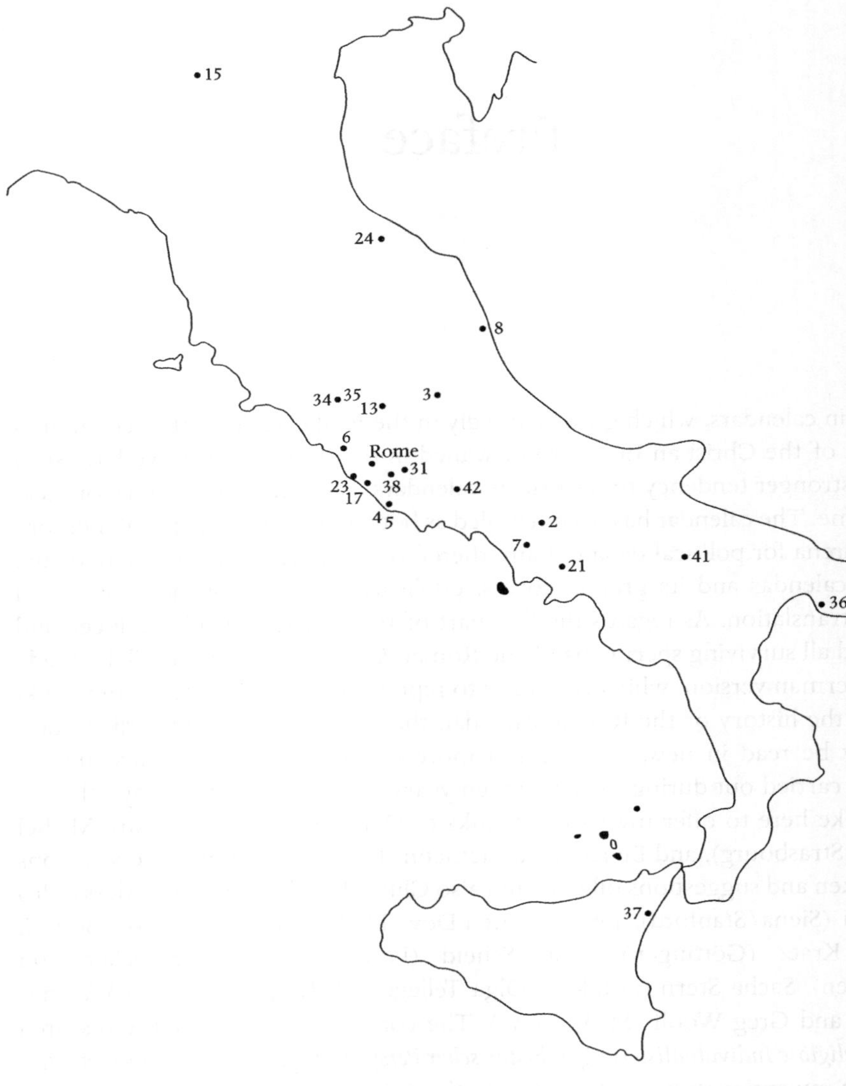

Map 1 The map shows the distribution of preserved calendars (or calendar fragments) of the *fasti* type from the first century BCE to the fifth century CE

Table 1 List of known copies of *fasti*

Name	*Origin*	*Date*	*Type*	*List of magistrates/ chronic*	*Heading attested*	*Characteristics*
1 F. aedis Concordiae	Rome	?	S	?	?	No small-letter entries? N for *nonae*
2 F. Allifani	Allifae	Tib.?	M	–	?	N; *feriae*; extended additions; very large
3 F. Amiterni	Amiternum	Tib.	M	+	?	Official copy? N; rich notes on imperial festivals; *feriae*
4 F. Antiates maiores	Antium	Late Rep.	P	+	–	Private; *Interkalaris; loedi*
5 F. Antiates ministrorum	Antium =	Tib.	M	+	+	College; N; short formulas for imperial festivals
6 F. Caeretani	Caere	Aug.	M	+	+	*Loedi;* continuous corrections
7 F. Capuani	Capua	Early Imp.	M	–?	?	Parapegma
8 F. Cuprenses	Cupra Maritima	Late Aug.	S	+	?	N; *feriae*
9 F. Esquilini	Rome	Aug.	M	+	+	*Loedi*
10 F. Fandozziani	Rome	Early Imp.	M	?	?	
11 F. Farnesiani	Rome?	Early Imp.	M	?	?	N; small entries
12 F. Filocali	Rome	AD 354	B	+	+	Lunar letters; no CD; *sen. leg.; dies aegyptiaci*
13 F. Foronovani	Forum novum	Tib.?	S	–?	+	College? HL
14 F. fratrum Arvalium	Rome	Early Aug.	M	+	+	Priestly college; *feriae;* continuous corrections

(*Continued*)

Table 1 *(Continued)*

Name	*Origin*	*Date*	*Type*	*List of magistrates/chronic*	*Heading attested*	*Characteristics*
15 F. Guidizzolenses	Guidizzola	Imp.	S	–	–	Only scheme of data; *feriale*
16 F. insulae Tiberinae	Rome	Early Imp.	M	?	?	N
17 F. Lanuvini	Lanuvium	Second century?	M	?	?	N; no CD/NL; *sen. leg.*; very small
18 F. Lateranenses	Rome	Early Aug.?	M	?	?	
19 F. Maffeiani	Rome	Aug.	M	–	+	College? hardly any small entries
20 F. magistrorum vici	Rome	Aug.	M	+	+	College; in a *Schola*? list of *magistri*
21 F. Nolani	Nola	Aug.	M	?	?	HL; few additions
22 F. Oppiani	Rome	Aug.	M	?	?	Few small entries
23 F. Ostienses	Ostia	Aug.	M	+	?	In colony; wordy notes on *feriae*
24 F. Palatii Urbinatis	Urbinum Metaur.	Aug.?	M	?	?	Extended entries
25 F. Paulini	Rome	Aug.	M	?	?	
26 F. Pinciani	Rome	Early Aug.	M	+	?	
27 F. Pighiani	Rome?	Caligula	S	–	?	*NF*; no abbreviations for festivals
28 F. plateae M. Fanti	Rome	Early Aug.	P	?	?	
29 F. Polemii Silvii	Gaul	AD 448/9	B	+	+	N; no NL/CD; prognostics; historical notes

30 F. porticus	Rome	Second/third century	P	–	–	Private *porticus*; images of months; no NL/CD
31 F. Praenestini	Praeneste	Late Aug.	M	+	–	Verrius Flaccus; commentary; N; notes on local festivals; very large
32 F. Quirinales	Rome	Early Imp.	M	?	?	
33 F. Sabini	Regio Sabina	Aug.	M	?	?	HL; no CD for Ides; N below DO
34 F. Sorrinenses maiores	Sorrina	Aug./Tib.	M	?	?	N
35 F. Sorrinenses minores	Sorrina	Claud.?	M	–?	?	Two months above each other; *sen. leg.*; *E* instead of *EN*
36 F. Tarentini	Tarentum?	Early Imp.	M	?	?	
37 F. Tauromenitani	Tauromenium	Tib.	M	+	+	College; N
38 F. Tusculani	Tusculum	Aug./Tib.?	M	?	?	No CD on DO/festivals; N below DO; small
39 F. Vallenses	Rome	Late Aug.?	M	+	+	Collegium; no CD on DO; many additions
40 F. Vaticani	Rome	Tib.	M	?	?	N; *feriae*
41 F. Venusini	Venusia	Late Aug.	S	+	+	In municipal town; N; signs of the *zodiacus*; rich additions
42 F. Verulani	Verulae	Tib.	M	?	–	N; large

(*Continued*)

Table 1 *(Continued)*

Name	*Origin*	*Date*	*Type*	*List of magistrates/ chronic*	*Heading attested*	*Characteristics*
43 F. Viae Ardeatinae	Rome	Aug.	M	?	?	
44 F. Viae dei Serpenti	Rome	Tib.	M	?	?	N; lines below date
45 F. Viae Graziosa	Rome	Tib.	P	?	?	N following DO; indications of place for Mens and Vestalia
46 F. Viae Lanza	Rome	Early Imp.	M	?	?	
47 F. Viae Principe Amadeo	Rome	Early Imp.	M	?	?	DO to the very left
48 F. Viae Tiburtinae	Rome	Early Imp.	M	?	–	
49 F. Vici Iugarii	Rome	Tib.?	M	?	?	N
50 F. villae Maxentii	Rome	Aug.?	S	?	?	

B: Book. P: mural Painting. M: Marble. S: Stone (lime). N: Numerals indicating distance to next Kalends etc. CD: Character of the Day. DO: Days of Orientiations. HL: Hebdomadal Letters. NL: Nundinal Letters.

1

Time's Social Dimension

Calendars have been invented in virtually every culture, for social life in a community requires structures defined by time as well as by space: in respect of each of these dimensions, chance circumstance and convention are commonly appropriated, and referred to 'sacred places' and 'hallowed times'. The effect is to legitimize and simplify. In respect of calendars, it is rarely possible to distinguish between religious and technological history; this makes them susceptible, liable even, to revolutionary change. The history of the Roman calendar more than any other can be followed through many centuries and many such revolutions, during which it has absorbed or displaced other calendars, or, as in Gaul and Palestine, been instrumental in bringing them into being. The purpose of this book is to trace that history, from its discernible beginnings until, in the shape of the Julian Calendar, it was able to accommodate the requirements of a Christian conception of time. It is still in worldwide use today, in the little-changed form of the Gregorian Calendar.

But what is a calendar? The name given the Roman calendar was *fasti*, that is to say 'list of court sittings'. The Latin word *calendarium*, on the other hand, meant 'register of debts'; it referred to the *kalendae/calendae*, the first day of the month, when loans were given and interest payments fell due. Already in the Early Middle Ages, however, Isidore of Seville used the word *calendarium* in its modern sense of 'calendar'. Thus neither of the Latin terms refers to the reckoning of time in the abstract. A 'calendar' is a document, a graphic, a text, with a particular look and a particular function. The reckoning of time is one aspect of such calendars, and calls for constant updating. Most extant examples of historical calendars contain amendments, additions, and deletions. Today too, most calendars are produced in forms that invite alteration and individualization. The resulting entries can change the face of a calendar, make of it a commentary or

---

The Roman Calendar from Numa to Constantine: Time, History and the Fasti, First Edition. Jörg Rüpke.

a historiographical work, complement it with specific texts and lists, or reduce it to a digest of locally relevant data. It is in the investigation of these different genres, and the inquiry into their communicative functions and the social circumstances of their creation, that the excitement of calendar histories resides.

Current sociological research provides a basis for understanding the problems facing a society in this context. Pride of place must belong to PITRIM A. SOROKIN and ROBERT K. MERTON, who, in 1937, on the basis of a few observations made by EMILE DURKHEIM,[1] provided a first, brief but far-reaching analysis of the concept of time. Time is a social construct, to be understood not in an astronomical, quantitative sense, but socially and qualitatively.[2] Systems for the reckoning of time, calendars among them, are founded on the need for coordination experienced by societies as they become more diverse;[3] thus the first calendars are found in towns and cities.[4] Religious festivals are only one of the elements recorded in such systems; astronomical phenomena are only occasionally called in aid, and then they are treated conventionally rather than strictly empirically.[5] This 'social time' is characterized by an absence of homogeneity: its continuity is interrupted by significant intervals whose own identification as unified entities remains contingent, and different emphasis is typically given to periods of time of equal length.[6]

The work of NORBERT ELIAS leads to similar outcomes. He brought together some important reflections on the calendar in a long essay entitled *An Essay on Time*.[7] His central idea is that of a gradually developing 'self-control of people in conformity with time', which becomes 'the symbol of an inescapable and all-embracing constraint'.[8] Time is a social institution with a coordinative and integrative function; only at a very late point in social development does the idea of 'physical time' emerge, and it is in this process of differentiation that the main interest lies.[9] As, in ELIAS' view, the symbols of time are subject to the pressure of 'object adequacy', the history of the calendar is linked in a particular way to the history of the measurement of time by observation of natural objects, and to the concept of time that emerges from that process: 'On a small scale, the development of that calendar is a good example of the long-term continuity characteristic of the development of human knowledge and connected aspects of human societies.'[10] In face of the radical dichotomy between social and physical time

1 DURKHEIM 1915: 11 n. 1: 'But what the category of time expresses is a time common to the group, a social time, so to speak. In itself it is a veritable social institution.'

2 SOROKIN AND MERTON 1937: 621–3.

3 See also SOROKIN 1943: 172, 181.

4 Sorokin 1943: 170.

5 SOROKIN AND MERTON 1937: 620–1, 624–6.

6 SOROKIN 1943: 183–4. In his ethnological remarks, the writer relies extensively on NILSSON 1920.

7 ELIAS 2007.

8 Elias 2007: 18.

9 Elias 2007: 10, 44, 100. Similarly SOROKIN 1943: 186–97.

10 ELIAS 2007: 157.

asserted by SOROKIN, ELIAS seeks to reconcile the two aspects. Of use in the analysis of actual calendars is his replacement of the romantic paradigm of an original unity of personal, socio-economic, and religious time, traces of which are supposedly preserved in the calendar, with the notion of a growing demand for systematization in societies of increasing complexity.

The momentum of this demand for coordination is highly relevant to the process ELIAS describes. 'The more complex social systems now become, the more pronounced the prominence of abstract concepts and structures independent of external events in the cultural construct of time. Time is no longer perceived as a sequence of events, as alterations in natural phenomena, but construed as a linear succession of instants that no longer mediate meaning and expectations on the basis of connection with a physical occurrence; a continuing example in the Gregorian Calendar is, for example, Sunday, devoted to the glory of God. Time becomes event-neutral.'[11] Only in this way can calendar-related constructs serve a society's manifold requirements for synchronization and the management of expectations. The same set of requirements produces a demand for the calendar's omnipresence, so that different versions, such as pocket calendars, wall calendars and tear-off calendars may emerge; as is evident from both modern and ancient examples, these different kinds of calendar by no means necessarily coincide in the coverage they offer.

The Roman calendar itself provides an instance of the problem of objectivity in the structuring of time. In analysing actual societies, we may be confronted with competing calendars, or situations where the relationships between mutually associated calendars are in some other way unclear. In some cases, the difficulty may lie in the fact that no written calendar exists. PIERRE BOURDIEU encountered problems such as these in his work among the Kabyles of the North-African Rif, when he attempted to produce a unified calendar from data provided by members of this ethnic group.[12] BOURDIEU advises caution when interpreting the resulting synopsis: prior account must be taken of differing individual perspectives and local[13] traditions.

Such instances can easily be augmented from everyday experience. Much calendar-related knowledge originates in quite disparate areas of life that, in the normal run of events, are not at all interrelated. Bluebell time, the strawberry season, Wimbledon, Indian summer, the autumn break, are all familiar concepts to us: but what dates[14] would we give to an ethnologist who wanted to reconstruct our calendar? Should he provide for such periods to overlap, or do they succeed one another, and are there gaps? Our ethnologist would most probably seek to

11 HEINEMANN AND LUDES 1978: 221. See also NASSEHI 1993: 332ff.

12 BOURDIEU 1990: 200–70. In the original French, the chapter is entitled *Le démon de l'analogie*, in the English translation, *Irresistible Analogy*.

13 KUBITSCHEK 1915 published ancient concordances of various local calendars ('hemerologies'); see SAMUEL 1972: 171–8; briefly 1988: 394.

14 On this problem see also TURTON AND RUGGLES 1978.

reconstruct a continuous year, so that the resulting calendar, in attempting to coordinate periods that are in fact discontinuous, would be a fiction that no individual would entirely acknowledge as 'their' calendar.

The resulting problems of recognition can easily be envisaged. It is hard to believe that the reconstructed calendar would resemble one of the genres familiar to us, and we would accordingly be unwilling to be bound by it. This binding character and the fact of publication are mutually dependent. The beginning of summer is astronomically determined, published, and thus unambiguously defined, whether or not it rains on 21 June. Even the beginning of the strawberry season does not depend solely on nature; it is determined in a discourse that includes classified advertisements, placards, and word of mouth. In the simplest as in the most complex societies, the natural benchmarks of time need social clarification and definition.[15] On the other hand, not every binding date is widely publicized a long time in advance; publication may be restricted to the circle of those immediately affected, and may not occur until immediately prior to the event, or even at a later time.

These few excursions into our everyday relationship with 'calendars' go to show that a historical analysis of such phenomena is not the same thing as a history of astronomy in the cultures in question. It is not the calendar as such that merits our study, but its mediated forms, and the ways they function in society, and in its institutions and subordinate groups. The purpose of this book is to submit the calendars developed in the city of Rome from its earliest period to that kind of attention. After an initial analysis of extant examples of calendars (Ch. 2) it will therefore be necessary, in the absence of archaeological evidence or authentic ancient testimony, to reconstruct the earliest Roman calendar systems on the basis of Late Republican and Imperial-Period accounts and theories regarding that earlier history. Especially relevant here is the reconstruction of the lunisolar, so-called pre-Republican calendar (Ch. 3), which forms the essential background to an understanding of the significance of the mid-Republican calendar reform (Ch. 4). The key to the question of the 'official version' and its religious function (Ch. 5) lies in the publication of the *fasti* by Cn. Flavius at the end of the fourth century BCE. The sections that follow are designed to substantiate further the conclusions arrived at so far. They deal with the problem of pontifical intercalation (Ch. 6), the brief entries, the dedication days, and the historical notes (Ch. 7). How did these arrive in the calendar? But, above all: into which calendar did they arrive? Other questions to be answered here concern the actual graphic models for the explosion of calendar production in the Augustan age, and the conception of the *fasti* that provided the basis for the Augustan project (Ch. 8).

The remainder of the historical survey is devoted to the history of the Imperial-Period calendar (Ch. 9). It inquires into the breaks in the tradition: first, why did marble calendars remain confined to the Early Principate, in contrast to all other

[15] EVANS-PRITCHARD (1939: 199–202).

epigraphic genres, whose incidence continued to increase until the Severan Period, at the beginning of the third century? The final stage covered is the encounter in Rome between Christianity, with its Jewish calendar tradition, and the Julian calendar: here can be seen in sharp focus the capacity of calendar systems to embrace, accommodate, and resist change.

2

Observations on the Roman *fasti*

2.1 A Republican Version

Before embarking upon any interpretation of a particular calendar or set of data, we must familiarize ourselves with the foundations on which it is based. In the present case, the Roman conception of the calendar is exemplified by a reconstruction of the Fasti Antiates (Figure 1), which survive in fragmentary form, and are on display today in the Museo Nazionale at the Palazzo Massimo in Rome. The year consists of twelve months; these are represented in as many columns, and begin with January. The names of the numeric months – September, October, November, December – are those still in use to this day; considered in combination with certain rituals, they indicate that, on 1 or 15 March, there was a break in the year that also had a bearing on the calendar: however, competing New Years are not an unusual phenomenon, and, so far as the calendar was concerned, the Republic may have known no other beginning to the year except January. On the far right, in this calendar from the 60s or 50s BCE, thus predating the Julian reform, can be seen the column of a thirteenth month. This is the 'leap month', called *Interkalaris*.

Four textual components, arranged partly in columns, can be discerned within the individual months. On the far left are serial letters from *A* to *H*, beginning on 1 January with *A*. These 'nundinal letters' mark the period of the eight-day Roman week (*nundinae*). In practice, these letters functioned like the capitals still used by the Christian calendar into modern times for Sunday. Of course, the week does not divide exactly into the year: this applies to the seven-day week of the Julian-Gregorian year as well as the eight-day week of the Roman calendar. Thus each year, then as now, the days of the week shift in relation to the date. If 29 December,

The Roman Calendar from Numa to Constantine: Time, History and the Fasti, First Edition. Jörg Rüpke.

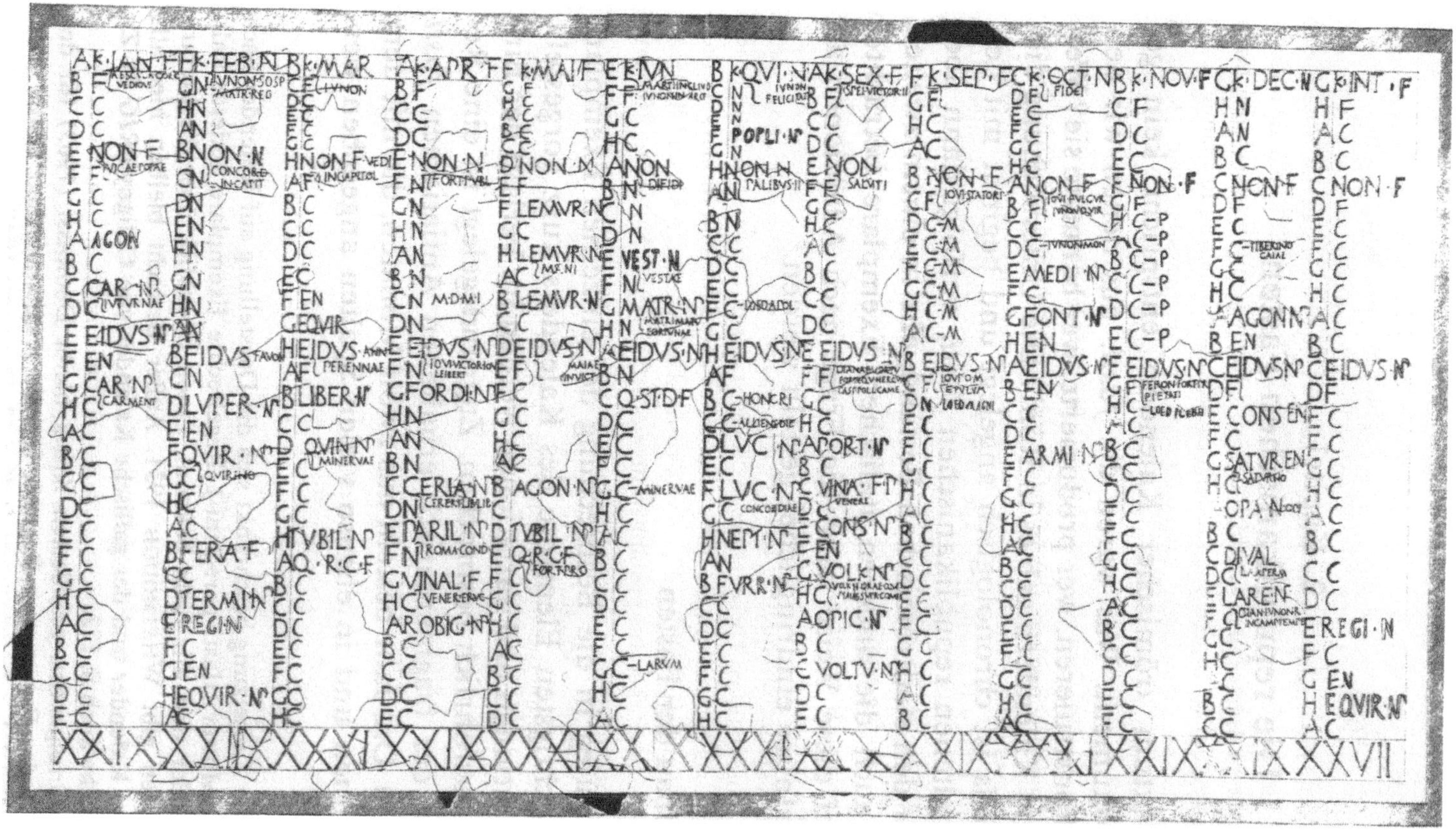

Figure 1 The Fasti Antiates maiores, in the form of a wall painting from a private building in the Latin city of Antium, represent the oldest preserved specimen of the Roman *fasti*, and the only one dating from before the reform of Gaius Julius Caesar in 46 BCE

marked with a *C*, is the last day of the Republican year, and the third day of a 'week' beginning with *A* (thus the supposed market-day letter for this putative year), then the fourth day of this 'week', 1 January, again bears an *A*. Consequently, the ninth day of the continuing sequence in this 'perpetual' calendar, does not fall again on an *A*, but on an *F*, which becomes the new market-day letter for the year now beginning. Unlike our modern pocket calendars, the Roman *fasti* were not conceived as 'throw-away' calendars, even in their portable form as papyrus rolls. All that changed from year to year was the identity of the letter designating market days.

To give a structure to each month, and for the identification of dates, three days were specially marked: the *Kalendae* at the month's beginning, the *Nones* on its fifth or seventh day, and – always on the ninth day following, hence the name *Nonae* – the *Ides* in the middle of the month, on the thirteenth or fifteenth day. These days were the basis for dating, dates being reckoned backwards from them. By Roman inclusive reckoning, 3 January is then the 'third day before the Nones of January', and 28 January (in a month of twenty-nine days) is the 'third day before the Kalends of February'. Some calendars give these numeric differences in a second column.

The next column frequently features a *C*, occasionally an *N (NP)*, an *F*, or one of a few abbreviations. These letters mark the status of the day, the day's juridical character. Decision-making people's assemblies could be held on days marked *C* for *comitia*; *F*, for *fas* ('allowed'), marks those days when cases could be opened before the *praetor*: this was prohibited on *dies nefasti* (letter *N*). About fifty days in the year, especially those marked *NP*,[1] bear multi-lettered abbreviations resembling the *NON* of the *Nones* and the *EIDVS* of the *Ides*. These abbreviations denote feast days, without doubt the most interesting element of the calendar. Finally, the words or abbreviations written in smaller letters feature the names of gods, and provide information on the foundation days of temples.

2.2 Forms and Functions

Apart from the two Late Antique book calendars, extant *fasti* belong to two genres: wall paintings and inscriptions. The surviving specimens are highly unequally distributed between the two genres.

The first group is represented by four specimens (panel calendars) in all (see Map 1 and Table 1). They are the Republican Fasti Antiates maiores, and from the city of Rome the Augustan Fasti plateae Manfredo Fanti, the Fasti Viae Graziosa from the time of Caligula, and the illustrated Fasti Porticus, which should be assigned to the Late Antonine-Severan period. The group covers a period of some

[1] See Ch. 5.2.

two hundred and fifty years, carried forward for a further two hundred and fifty years only by the book calendars. The design of these wall calendars does not differ from that of versions in stone. This also applies to their size. They take up from one and a half to three square metres; with the consul lists, the Fasti Antiates attain four and a half square metres. The enormous frescos of the Fasti Porticus demonstrate besides that illustrations could accompany *fasti* in wall-paintings and books.

The two later calendar paintings can safely be assigned to a private context; the same is suggested in the case of the oldest specimens by the find context. It appears legitimate to conclude that, in the Imperial Period at least, wall calendars commissioned by individuals for their private dwellings took the form of paintings. In the case of colleges and contexts that were more official still, this style of execution was seen as inadequate. The conclusion does not hold for the Republic, where we know of no competing stone calendars: even the *fasti* of M. Fulvius Nobilior in the temple of Hercules Musarum[2] were probably painted.

Inscriptions make up the overwhelming majority of extant calendars. This may be due to differing survival rates for ancient wall-paintings and stone inscriptions. On the other hand, the distribution over time is significant. Most specimens by far are from the Augustan-Tiberian period; only isolated instances of *fasti* occur later, and those deviate increasingly in form. No calendar from Rome itself can be dated with certainty to the period after Caligula. Outside Rome, if we disregard the 'Fasti' Guidizzolenses, only the Fasti Lanuvini can be dated to the second century of the Current Era, and then only tentatively. Stone calendars are thus distributed over scarcely more than one hundred and fifty years, more than 90 per cent of them occurring in the first half of this period.

The dilapidated state of surviving specimens has long closed our eyes to the fact that calendars must be counted among the most important genres of inscription from the time of Augustus and Tiberius. They nearly always comprise a large-scale mural decoration executed in marble. The average size in Rome is about two square metres, and outside Rome about three, with a considerable degree of variation; each of these categories includes some twenty specimens. They commonly measure eighty centimetres high by one and a half metres wide; several later Italian calendars measure more than one metre high and about four metres wide.[3] Unusually large examples are, in Rome, the Fasti fratrum Arvalium, and in Latium the Fasti Praenestini; the latter cover twelve square metres, while the former are incorporated in the ongoing, monster inscription of the Acta Arvalia. The difference between these and many other specimens may, however, not be as great as appears, as calendars have survived only in small fragments, and we can often not say whether they were accompanied by magistrate lists, which can increase the size of the inscription considerably.

2 See Ch. 7.

3 See the brief indications in DEGRASSI 1963: xxv.

The Fasti Maffeiani, even though today no longer complete as an inscription, count among the best-preserved calendars. At the same time, at ninety centimetres wide and just seventy high, this calendar is one of the smallest specimens, and it is possible to detect in it some correspondence between size and density of information. As regards additions written in small characters, it is very restrained. Large calendars too can appear 'meagre', but it is in those *fasti* that are the largest in area that we find the most comprehensive information. In the size of characters used, of course, calendars cannot compete with immense building inscriptions and Imperial-Period honorific inscriptions; but it is worth noting that, in a complete calendar, solely the portion written in large letters amounted to some one thousand characters. Additions were normally of the same order of size, but could sometimes be much more extensive:[4] if magistrate lists were included, this multiplied the number of characters. Seen in this light, the *fasti* exceed most genres of Latin inscription in density of content. They are surpassed in this only by such exceptional epigraphic documents as *acta* and *commentarii* (e.g. the Acta Arvalia), legal digests (if we think of municipal statutes), and the Res gestae divi Augusti.

Fasti change enormously after the mid first century CE. Overall, however, the consistency of content is astonishing. No example, whether painted or incised, can seriously be treated as an 'engagement diary' for an individual or group. At most, a calendar might be individualized by selectivity in respect of the small additions included, but even this practice is confined to the tiniest variations.[5] Divergences occur, if at all, at the margins: weekly letters, whether nundinal, for the Roman week of eight days, or hebdomadal, for the planetary week of seven days, are added to the left of the core information, while 'additions' occur at the right-hand margin.

There are no other 'blank spaces' available. Values indicating intervals before the next 'orientation' day, and inserted between nundinal letters and daily characters or abbreviations for feast days, tend to appear in more recent calendars. They form a secondary element, an explicit statement of information already implied in the way the months are arranged in their columns. This becomes highly important when we come to assessing the function and motivation of the stone calendars. As the Julian calendar reform bore mainly on the lengths of months, in eight out of twelve months half of the values marking intervals changed: all days after the Ides in months with less than thirty-one pre-Julian days were affected. If the epigraphic *fasti* had been created in reaction to the practical demands of coping with the new calendar, such values would have had to feature strongly from the beginning.

[4] The figure of 2,500 'small' letters may well have been exceeded in the case of the Fasti Amiterni. Taking as a basis the extent of the text of the Acta Arvalia entries for individual years, and the associated time required *scripturae causa*, documented by the dates of the piacular sacrifice, SCHEID (1990: 86-8) arrives at a rate of 180 to 300 letters per day. Modern-day stonemasons, working manually, achieve a work-rate of some 250 letters per day (GORDON 1983: 32).

[5] Especially in the Fasti Maffeiani and *Antiates ministrorum*.

The inexactitude of the calendars appears astonishing. For one thing, there are glaring errors. It is not only that the reasons for feast days are missing: characters of days are misplaced; *NP* days, one might think the most important information in the calendar, are often not marked as such. The difficulty of many datings results precisely from the fact that, in respect of Imperial feast days, recent *feriae* decrees are marked, while others, dating back some time, are missing for no apparent reason.[6] But still more unsettling than the errors is the apparent optional character of the entries. The abbreviations are by no means canonic, but can comprise three, or four, or as easily six letters. Additional entries may be absent, abbreviated, or written out in full. It is difficult to imagine that such formal decisions were made with no plan, and, for the most part, it is in fact possible to detect a system behind them. But what cannot be discerned is the deeper reasons. In some places, but not many, it is possible to suspect similar 'recensions' in different *fasti*, and thus common sources: but no *stemma* emerges. It is probable that scarcely any inscription ever served as a source for others. The true bearers of the tradition were the book calendars, whose earliest extant specimen is Furius Filocalus' sumptuous version from Late Antiquity: or, more precisely, medieval copies of it.

Particularly striking is the treatment of temple-foundation days. Individual *fasti* differ considerably in the extent to which they mark these days. This is never due to any special interest in respect of content: no *municipium*, for example, chooses only the gods represented by temples within its confines or in its vicinity. Where it is possible to compare several extant entries for one day, it becomes apparent that only the first instances from a longer list are included, to the extent that space allows; or the first and third of four items, with no apparent reason. There is no correlation of content with the social or geographic source of the various specimens. There is no difference, in this respect, between Italian (and even Sicilian) calendars and those from the city of Rome.[7] The lack of correlation between content and context of installation also applies to the size of location, extending from *vici* and *praefecturae* to important *municipia*: village calendars do not differ from calendars in big cities.

The calendar's content remains constant. If we turn to inscriptions outside Rome, the nearer environs dominate: Latium, Southern Etruria, the Sabine country as far as Umbria, Campania, and Samnium. It has proved possible to assign more distant specimens to Roman colonies such as Apulian Venusia. The Fasti Tauromenitani even come from a Roman colony outside Italy, in the province of Sicilia. They all, without exception, list exclusively feasts and temple dedications celebrated in the city of Rome, some of them capable of being celebrated only there.

[6] DEGRASSI 1963: xxii.

[7] Cf. HÖLSCHER (1980: 321), who stresses the general point that the imagery of the Roman Imperial Period sets itself apart from the no longer generally understood idiosyncrasies of Late Republican symbolism by the quasi-ritual use of predetermined motifs (cf. Ch. 5.4, 5.5).

The hypothesis that comes most readily to mind is centralization: essentially, local calendars are displaced by those of the city of Rome; local cult, juridical, and political functions are transferred from the periphery to the centre.

Is this so? The calendrical landscape of Republican Italy resembled that of Greece, with a variety of local, lunisolar calendars.[8] Testimony of the Etruscan system is provided by the *Tabula Capuana* and the mummy wrappings from Agram; Italic calendars are known to us from antiquarian sources, a few dates only from inscriptions.[9] In assessing the antiquarian evidence, difficulties arise in respect not only of the social and chronological origins of the calendar described, but of the degree of confidence that can be placed in the information given. Strong pressure on entirely deviant calendar systems may be assumed for the period after the Latin wars. As the Greek example shows, differing names of months do not necessarily indicate differing systems for the reckoning of time: a *Flusaris* in Furfo may easily function as an indigenous label for a 'Roman' month. Similarly, New Year's Day was still being altered in the Augustan period,[10] and so differing dates do not necessarily indicate a fundamentally different calendar system. A 'Roman' system cannot be excluded even in the case of the mummy wrappings from Agram (technically a *feriale* rather than a calendar).[11] I surround 'Roman' with inverted commas, because the Roman calendar itself should be considered in an Italic context. Thus Etruscan-Latin traits should be assumed for the pre-Republican calendar, and external influences cannot be excluded even for the calendar reform that resulted in the pre-Julian calendar.

How the situation stood in the first century BCE can only be surmised. The will for integration, radiating from the Italic peoples, and the pressure for Romanization after the Social War, together with the decline of regional dialects,[12] may also have engulfed any calendar systems that survived. Throughout the Fasti Praenestini, Verrius speaks of the Latin month-names entirely in the past tense. What was the role played by our stone calendars in this process? They would seem, in any event, to have come too late to be active instruments. In most important areas, the Civil War may already have brought the process to an end by the last third of the century: that is to say, before the production of the first marble calendar.[13] On the other hand, the Fasti Tauromenitani, with their reference to the local Greek calendar, show that the installation of *fasti* did not necessarily presuppose the adoption of the system for the reckoning of time on which they were based. However, too much must not be made of this exception, which is in any case problematic: the geographic distribution concentrated in Central Italy, and the favouring of

[8] SAMUEL 1972; HANNAH 2005.

[9] See WHATMOUGH 1931. *ILLRP* 508 = *ILS* 4906 provides double dating for a temple dedication in Sabine Furfo, with the month-name *Flusaris*. On the Iguvine Tablets see BORGEAUD 1982.

[10] Suet. *Aug.* 59.

[11] PALLOTTINO 1985: 324.

[12] See the overview in HARRIS 1989: 178; in more detail SIMONE 1980.

[13] See PALLOTTINO 1991: 151-69.

small towns whose own traditions were less well-established, suggest progressive use of the Roman system.

In the East of the Empire, as in islands of Greek culture in the West, Greek calendars or Imperial-Period adaptations of the Julian calendar functioned at a local level (not in the Imperial administration and the army).[14] With the above-mentioned exception of the Fasti Tauromenitani, we know of no Roman *fasti* from cities in these areas. With the exception of the Gaulish calendar from Coligny,[15] evidence of calendars is lacking in the Western provinces; what indications there are lead one to conclude that, as in pre-Augustan Rome, there was no tradition of epigraphic calendars. The Roman, and above all the Julian, calendar was certainly used for administrative purposes, but it is uncertain how far it remained a mere superficial element, with competition from other calendars persisting, or became established, and, taken for granted at all social levels, was not displayed on material that could physically survive.

There can be no question of a centralizing function for the *fasti* in connection with the introduction of a new method of reckoning time. The same negative conclusion applies in other areas. Centralization would have to be combined with demonstrable functional changes. There is no evidence of such changes in the area of religion. No regulating function in respect of local cults can be effected by a calendar that takes no account whatsoever of such cults: changes in the dates of local feast days to fit in with the calendar of the city of Rome would have left no trace of any kind. Even to read the *fasti* as an invitation to complement local festivities with a visit to Rome appears far-fetched. The basis for such a choice did not exist. Journeys were much too arduous for such encouragement to have any effect outside Rome's immediate vicinity.

But, as information about court sittings in Rome? To avoid incurring an unnecessary burden on the travel budget by kicking one's heels for several *nefas* days? Here too, the situation on the ground did not support such a hypothesis. Even in the less urbanized localities, the prefectural system provided adequate legal services. The postal system stood available, and, in case of doubt, it would have been a matter of complying with court sittings rather than choosing them. What is more, much legal business was not at all affected by the *nefas* sanction imposed by *N* days, which related to the formal opening of a case before a praetor, the *legis actiones*.[16] Finally, if the hypothesis were true, more care would have been taken to ensure the precision of the information concerning the characters of days.

Did these inscriptions, installed in such numbers, and certainly at no little expense, have any function whatsoever? To come closer to answering this question, we must entirely change our frame of reference. Form (the *fasti*) must be

[14] SAMUEL 1972: 171-88.

[15] The best edition now *RIG* 3; more recently OLMSTED 1992.

[16] MICHELS 1967: 68; TELLEGEN-COUPERUS 2006.

scrupulously separated from content (the 'calendar'). Of aid in the accomplishment of this task will be the striking character of the inscriptions' distribution over time.

2.3 The *fasti* and the Birth of Augustan Epigraphy

In his book *The Power of Images in the Age of Augustus*, PAUL ZANKER describes the rise of a specifically Augustan stylistic and formal language, one might even say an Augustan architecture, which was both a consequence of particular combinations of political circumstances and the bearer of a political programme of religious and moral renewal. In a few decades, Republican Rome acquired a central core of marble,[17] and within a century had been transformed into the marble city of the Flavian period.[18] GÉZA ALFÖLDY, with all his reservations about the implied public image of Augustus, and the focus on the person of the emperor, has taken up ZANKER's interpretive method, and applied it to the rise of the epigraphic culture of the Imperial Period.[19] In genre after genre, he has demonstrated the defining influence exerted by the emerging Augustan monarchy on the forms of building inscriptions, honorific inscriptions, milestones, and gravestones, and the formulas used in them. Neither of these authors gives any attention to the surviving calendars, pitiful fragments that they are; but, in exemplary fashion, they illustrate the process I have described. These great, lavish marble tablets represent an epigraphic genre that is nowhere to be found in the previous period; in the age of Augustus, however, they spread like wildfire across a wide area of Italy, leaving in Rome alone some twenty identifiable specimens.

The thesis I am advocating is as follows: in the context of the emerging epigraphic culture of the Augustan Period, the *fasti* too were 'discovered' as potentially decorative inscriptions, and, chiselled in marble, soon became widespread. A religious function was also part of the process, insofar as this form of calendar was also understood as an emblem of religious renewal under Augustus. The question as to how far this religious dimension can be distinguished from the reference to religious and cultural renewal under Augustus to be found in almost all public buildings of the period, together with their decorative motifs, must await later detailed analysis.

We shall begin with the external elements of the phenomenon. The visual appearance of the Augustan *fasti* corresponds to Republican examples. This is confirmed by the Antium calendar. There can be no doubt that suitable models were available not only in the Antium villa, but also in Rome. There is nothing to

[17] ZANKER 1990: 28, 104, 139, 257, 308-11.

[18] BOYLE, DOMINIK 2003.

[19] ALFÖLDY 1991; s.a. ECK 1984.

suggest the existence of epigraphic exemplars. Macrobius attests to graphic representations, mentioning *fasti* in the temple of Hercules Musarum.[20] Cicero and Ovid both indicate that portable calendars in the form of papyrus rolls were common. The same transition to another medium also applied in the case of magistrate lists, which were part and parcel of this same process, and were likewise referred to as *fasti*.[21] There is no indication of epigraphic versions under the Republic.

The calendar's lack of an epigraphic exemplar provided inscriptions with an unusual advantage: freedom from context. Ordinarily, inscriptions comprise devices or captions; they derive their meaning to a great extent from the object with which they are associated, and that they explain: structures or restorations; a funerary monument or sarcophagus; a statue, to whose base the text is affixed. Graffiti, on the other hand, are in this sense 'context-free', their connection with the object on which they are written being contingent, without internal necessity, even though the choice of location may have direct relevance. This, to some extent, explains the different social contexts of inscriptions and graffiti: the object decorated with an inscription costs more money than the inscription itself, while, for the writer of graffiti, his 'canvas' is cost-free. In the case of inscriptions too, the relation between the significance (and cost) of the text and the significance (and cost) of the object can vary. Building inscriptions and artists' signatures, which may or may not be present without essentially reducing the utility of the object, constitute one extreme; gravestones and milestones, whose size is materially governed by the size of the inscription, the other. The unwonted boom in milestones in the Augustan period[22] indicates awareness of this economical consideration. The cost of the inscription itself was, of course, exceeded by other costs that should not be underestimated, such as road-building or burial. Even a cenotaph presupposes the existence, somewhere, of a corpse.

It is scarcely conceivable that calendars were as economically priced as milestones or headstones. They were not, however, dependent on a particular context. The object essentially comprised the calendar inscription itself, which could in turn, perhaps in the form of a heading, become the object of another epigraphic record.[23] This function is expressed by the form in which they were normally manufactured: thin marble panels, capable of being used to decorate buildings. As to the precise manner by which they were affixed, the archaeological record is silent. In no instance have *fasti* panels been found in conjunction with elements of a substantial surrounding decoration, into which they may have been set,

[20] Macrob. *Sat.* 1.12.16.

[21] Cic. *Att.* 4.8*a*.2; Ov. *Fast.* 1.657.

[22] See Alföldy 1991: 299-302.

[23] See the Fasti Maffeiani and Vallenses. An inscription from Antium celebrates the installation of *fasti* (this cannot refer to the Fasti Antiates ministrorum). *CIL* 10.6679: *(5) [Aedem lapide quad]rat(o) et ostia et fastus de sua pecun(ia) fecer[unt].*

although there is often an integral defining border. With the exception of the Fasti fratrum Arvalium, *fasti* may have been individual decorative components rather than parts of a greater, still more lavish, marble decoration on an already existing building. In particular instances, a free-standing position might be chosen, either because no suitable existing building was available, or, as in the case of the *vicomagistri* and in Praeneste, in order to give the *fasti* prominence as a monument in their own right. All in all, they comprised a type of inscription suitable for prestige purposes, but available even to purses of a limited (although still substantial) size. It is not merely that calendars have been found in small and even tiny localities: they have been found predominantly in such places. Their patrons, where they can be identified, were far more likely to be small associations than great *civitates*.

Which brings us back to the question of the archetype, the initial Augustan inscription. The utter silence of the literary sources as regards a monument of Augustus or Agrippa that could have served both as an example and an impetus is significant not only intrinsically, but in the context of the above considerations. The Horologium can be excluded immediately: it cannot have been the model, either formally or chronologically. The Fasti Capitolini would be the next possibility, but they must be ruled out in respect of the calendar: the fact that, among the multitude of fragments, not one has been found that could be assigned to a calendar, permits us to conclude that no calendar was ever present. On the positive side, the existence of the triumphal list complete down to Romulus betrays a departure from the usual combination of calendar and magistrate list. The Doric Order chosen for the triumphators, contrasting with the Corinthian-style architectural context of the consul list, betrays awareness of the novelty of the combination, probably realized in 19/18 BCE in an Augustan arch on the south side of the temple to Divus Julius. There was, therefore, in the centre of Rome no Augustan monumental calendar that could have directly inspired all imitators: a significant finding. There was, however, an artistic model, and it has survived relatively intact. It was located at the fifth milestone on the *via Campana*.

The *lucus Deae Diae*, the cult site of the *fratres Arvales*, possesses in the Acta Arvalia one of the greatest complexes of Latin inscriptions. The Fasti fratrum Arvalium form part of that complex, and, according to JOHN SCHEID, the greatest expert on these particular *fasti*, Augustus reorganized the college in 29/28 BCE. The basic furniture of the site comprised, apart from the buildings, a calendar, lists of magistrates, and the first panels of *commentarii*.[24] This calendar therefore represents the earliest datable Augustan *fasti*, and the largest we know of in the city of Rome. The same applies to the consul lists and the *commentarii*. It most probably represents the first ever rendering of the three genres on stone in Rome: was this by pure chance?

[24] SCHEID 1998.

We know of the Arval brethren, their cult, and their founding myth, from numerous sources. Their existence during the Republic is confirmed by a reference in Varro.[25] The agrarian-orientated cult of the Imperial Period may be Republican, at least in its fundamentals. Although it is not possible to establish with any certainty the historical source of the myth referred to in Imperial-Period sources, the little evidence we have tends to support the hypothesis that it was an Early Augustan creation.[26] The central element for the purposes of our analysis, however, does not originate earlier than the Augustan period. Literary accounts, but above all the content of the Arval Acts, unmistakably indicate that Augustus' reorganization defined the brethren as a central element of the Imperial cult, as conceived by him at the beginning of the twenties BCE. This obscure Republican priesthood was transformed into the most socially prestigious college, with close links to the Imperial family.[27]

Against this background, what profile do the inscriptions present? Their position inside the grove, in a place that did not facilitate publicity (while not absolutely preventing it), was not entirely without parallel. Many honorific inscriptions and epitaphs were in locations that restricted viewers to a small circle: the house and its visitors, the observant family at the graveside;[28] but inscriptions of this size? The content, at least that of the *commentarii,* is, by virtue of its provenance, not designed for a broad public; even though, in general, no all-embracing arcane discipline was incumbent on the priesthoods. The prestigious furnishing of what was in a sense a new college is conceivable as a project of C. Caesar, to be celebrated in the near future as Augustus, and he must have been at least indirectly behind the decision. The elite of the Empire, who were to make up the reorganized college, themselves alone comprised a worthy and worthwhile public. This is not to deny the possibility that the inscription represented a formal offering to the divinity to be revered;[29] primarily, however, its purpose was to confront the new members of a previously obscure college with the significance attached to their task and to their archaic rituals.[30]

On the other hand, there were enough opportunities in Rome to make effective use of artistic and financial resources. We may perhaps come even closer to the patron's intentions if, to the idea of an exemplar, we add that of an experiment: the grove in its seclusion to function as a test-bed for Rome's first large-scale historical inscriptions, an unparalleled threefold crossing of genres, and the first epigraphic calendar (for which no Greek precedents were available). As regards

[25] Varro, *Ling.* 5.85: *Fratres Arvales dicti qui sacra publica faciunt propterea ut fruges ferant arva: a ferendo et arvis fratres arvales dicti. sunt qui a fratria dixerunt. fratria est graecum vocabulum partis hominum, ut <Ne>apoli etiam nunc.*

[26] For a full account see SCHEID 1975: 352-63; cf. SCHEID 1990: 695 n. 16.

[27] For an examination of this aspect see in particular the comprehensive prosopographic studies of SCHEID (1975, 1990a), and the detailed study in respect of the Flavian Period by SYME (1982). See also BEARD 1985.

[28] ECK 1984: 134-5.

[29] SCHEID 1992: 121.

[30] Cf. RÜPKE 2004.

the magistrate lists and records, Augustus must have felt that his experiment had succeeded. Ten years later, the Parthian Arch was probably decorated with colossal lists of consuls and triumphs, and, only a little later still, the protocols of the secular games of 17 BCE were inscribed in stone and bronze.

There is no knowing what success Augustus attached to the Arval calendar. The fact that the combination of elements in the Fasti Capitolini omitted the calendar requires explanation. There was evidently no perceived pressing need for making one entity of a two-part chronographic apparatus. The *princeps* had perhaps also realized that *ad hoc* enlargement could lead to significant empty spaces in a marble calendar, and a correspondingly unsatisfying overall aesthetic impression: a problem of monumentalization that prevented calendars from emulating the Res gestae, and covering entire temple walls. For projects of a more modest, less monumental nature, however, the effect of the Fasti fratrum Arvalium must have been convincing. The *fasti* were, in any case, familiar as wall-paintings and in book form, and so the general model provided by the inscriptions displayed on Rome's outskirts and at its centre had no need to wait for imitators. And *fasti* did not have to be ordered by the monarch.[31]

Splendid as the polished marble tablet may have been, this was not yet sufficient reason for hanging a large calendar on one's wall. Other motives were ready to hand. To anyone not wanting to be excluded from the vogue for inscriptions and foundations, a calendar that could be affixed to an already existing building offered certain economic advantages. The Arvals' initiative provided a kind of approval, and the supposed archaic character of the *fasti* no doubt played a positive role in view of the Augustan emphasis on *pietas*. We can only speculate as to further grounds: the calendar, perhaps, as a pictorial puzzle, inviting the viewer to resolve its abbreviations, but effective even without that understanding, and, like other Augustan symbols, capable of banalization as a decorative item;[32] a 'text' that had something to say even to the illiterate (and, universally, they constituted the majority of viewers),[33] and was easily recognizable by virtue of its form and structure. An additional attraction was the opportunity to set up on one's own account a document that possessed at least specious authority.

We must assume a religious, 'iconic' dimension to the way *fasti* were received by their public. The dedications, at least, refer without any doubt to cult matters: gods, temples, *sacrificia sollemnia*, and annual festivals. Frequently in calendars, feast-day abbreviations written in large are explained with a note assigning the *feriae* to a particular god. Such impressions did not exist in a vacuum; at the same period in Rome, temples were being restored and decorated in the grand style, and new ones built. Although this had happened in the past, in the substantial

[31] On those means of dissemination that rely on fashion and quality, and competition between their creators, rather than direction from the centre, see ZANKER 1990: 265ff., 1988: 26; ALFÖLDY 1991: 315-22.

[32] Cf. WALLACE-HADRILL 1990: 159.

[33] HARRIS 1989; BOWMAN 1991: 122; CORBIER 1991: 117.

construction programmes of such figures as Pompey and Caesar, now, under the patronage of a broader stratum of active commanders and magistrates, it entered more powerfully into the general consciousness. 'Religious renewal' assumed a central status in the political and politico-cultural programme of Caesar's heir. Increasingly, there was a perceived advantage in having oneself portrayed in religious roles.[34] Religious motifs, portrayals of gods, and mythological scenes, attest that 'many prosperous individuals evidently … wished to put images of the new religiosity on display'.[35] The motivation behind broadcasting a text capable of evoking not only this complex of factors, but also the great feast days mentioned in the text itself, appears easily understandable.

A second important area may be defined under the heading of the historical dimension, although with certain necessary reservations in respect of the calendar component. We cannot know to what extent the calendar's abbreviations were capable of evoking in the mind of an ordinary citizen the many historical legends and foundations listed by a Verrius or an Ovid,[36] or whether this man in the Roman street would have been capable of recognizing references in the calendar to Aeneas, Maezentius, Euander, Romulus, Numa, Tarquinius Superbus, and others. Although the relevant stories were recounted by mythographers such as Ovid, and, to a certain extent, probably an autonomous mythic tradition, the Augustan imagery made no substantial reference to them. Narrative friezes played a diminishing role.[37] It is not possible to discern a coherent 'ideological' or 'propagandist' line.

The historical dimension is more explicitly expressed in the magistrate lists, which were frequently displayed in direct context with the calendars. With consul lists that often went back decades, they provided an outline of a past that producers of *fasti* in the city of Rome, and, after the Social War, Italians living outside Rome, could alike comprehend as their own. The extension of the eponymous lists to include military occasions and suffect consulates attests to an understanding that extended beyond the realm of mere chronography. It was here that the form offered the greatest possibilities for diversification, above all in the opportunity to include documentation of local magistracies, whether of municipal officials or officers of associations. Here too lay the greatest scope for active identification with the centre, for establishing a parallel in local history, capable of being embedded in the history of the greater entity. If we are looking for a centralizing function for the *fasti*, it is here, in the 'centralization of the historical perspective', that we are most likely to find it.

Ovid reveals the third central theme in the introductory and dedicatory sections of his commentary on the *fasti*: the typically Augustan touch within the Augustan

[34] ZANKER 1990: 123-35.

[35] Zanker 1990: 267.

[36] See Ov. *Fast.* 1.7: *Sacra recognosces annalibus eruta priscis.* For an examination of the diverse strands of an antiquarian discourse see SCHEID 1992 and RÜPKE 2009b; for a fundamental study of the emergence of 'commentaries' on rituals VANSINA 1985: 10.

[37] ZANKER 1990: 206.

fasti. Thus, in the reworked opening, Germanicus' attention is drawn to the work's content:

> There you will find festivals that concern your house too:
> You will often read of your father, and of your grandfather.[38]

This is precisely paralleled by the original proem addressed to Augustus in the second book:

> And yet, Caesar, it is with a dutiful heart that I pay homage to
> Your name, and trace your renown through your titles.[39]

The reference to Augustus in the caption to the Fasti magistrorum vici constitutes the exception to the rule that factors intrinsic to the calendar are central here: not only the month-names *Iulius* and *Augustus*, but also, and quite particularly, the multitude of Imperial feast days, historical memorial days to victories, triumphs, and personal feast days of the Imperial house. Examination of the Fasti Praenestini, Fasti Amiterni, Fasti fratrum Arvalium and others reveals the extent of this theme.

At the same time, an important and much more traditional theme of the calendar must not be forgotten: the temple-foundation days. Investigations of the foundation dates have repeatedly revealed duplicate dates for individual temples. They can best be explained by discrepancies between the Republican foundation days and the dates of Augustan restorations. It is, however, by no means self-evident that such discrepancies should be present. A more likely strategy would have been to combine the rededications with the traditional foundation dates: but this would have meant that they could not have been identified in the calendar. The practice thus indicates a conscious policy to give maximum prominence to the restorations, and, in the final analysis, allow them to appear in the calendar as new foundations. If we add to this the tendency to combine the new dates in a significant way with feast days associated with the emperor's person and family, it appears that there is an undeniable predisposition to monarchize the calendar, even where the name of Augustus does not appear.

The project to give the calendar an Augustan gloss may be cited as a factor in the early phase of its spread, with one reservation: the references cited above are from Late Augustan or even Tiberian *fasti*. The stonemason responsible for the Fasti fratrum Arvalium added the Imperial feast days as something of an afterthought. These feast days came to dominate the visual aspect of the *fasti* only once the accumulation that had begun under Caesar had been consistently maintained, or even intensified, throughout Augustus' long reign: perhaps a further reason why Augustus did not want to fix that same visual aspect earlier, in the

[38] Ov. *Fast*. 1.9-10.

[39] Ov. *Fast*. 2.15-16.

context of the Fasti Capitolini. The month of *Sextilis* was not renamed until 8 BCE.[40] The accumulation of festivals may also rapidly have become a problem when the mass of old feast days threatened to obscure new additions: the pruning of excess feast days decreed (to himself) by the preceding incumbent thus became an expedient act for any new emperor at the beginning of his reign. At the same time, such an exercise was a suitable means for a newcomer to create a usable canvas to fill with his own festivals. We cannot tell from the available inscriptions whether all these developments represented entirely new departures, or merely perpetuated characteristics of the *fasti* that had already existed under the Republic: this problem too requires more extensive historical investigation.

2.4 The Question of the Archetype

Once already, we have asked whether it is possible to point to an archetype of the surviving calendars. Then, we confined the question to the artistic model that inaugurated the genre of Imperial-Period stone calendars. That question now arises again from an entirely pragmatic, religious-historical point of view: did there exist at the centre, in Rome, a model calendar from which the enormous number of amendments and corrections occasioned by Augustus' religious policy could have been taken? Where was this calendar situated, and what was its official status? And, finally: when was it created?

I am not aware of any serious discussion of this problem. *Communis opinio* may be represented by WISSOWA, when, on the basis of the archaeology as it stood in 1902 and 1912, he writes:

> From the period beginning approximately with the foundation of the Principate and ending with the emperor Claudius, we possess some twenty *copies* of the Roman state calendar designed for *use* in Rome and the municipal communities in Rome's vicinity; it is obvious that these copies all refer fundamentally to the same *official version*: considered in combination, they go to make up an almost seamless picture of the Roman *religious year*.[41]

To summarize the four key statements contained in the passage, the hypothesis is that the official religious calendar consisted of an original version in the city of Rome, from which copies were made, as with legal texts, and published with the force of binding rules in other localities.[42] What was the source of this archetypal

40 Dio Cass. 55.6.6. Macrobius provides us with the text of the Senate resolution, which was confirmed by a *plebiscitum (Sat.* 1.12.35).

41 WISSOWA 1912: 2 (1902: 2); the italics are mine (Rüpke). English version by the present translator.

42 See e.g. FOWLER 1899: 11: 'The rectification of the calendar by Julius, and the changes then introduced, brought about a multiplication of copies of the original one issued under the dictator's edict.'

version? On the basis of present knowledge, the answer might be that, if it was not published at the time of the Twelve Table laws, the calendar had certainly stood in the Forum since its publication by Gnaeus Flavius in 304. Initially it was only on wooden tablets, until the *pontifices* replaced these with stone *fasti*, continuing to add changes as they arose. But we will return to these problems in the chapters that follow.

Plausible as this reconstruction may appear, it does not bear closer inspection. Although we may pass over completely the regrettable fact that no evidence for the 'official version' can be found in the sources, internal contradictions remain. The tradition tells us that Flavius published the *fasti* without the agreement of the pontifical college. Are we to believe that the college resigned itself to the new set of circumstances, to the extent of making this provocative act the official basis for its future functions? While not impossible, this is, to say the least, improbable. Contradictions are also apparent in the 'seamless picture' WISSOWA conjures up: even when due regard is given to diachronically based differences, the discrepancies, not to say errors in the different versions of the *fasti* remain astonishingly great, and disqualify them as technical tools.[43] When subjected to such investigation, the hypothesis is entirely destroyed: no question of 'use' in any practical sense; instead of 'copies' a craft tradition of decorative tablets; instead of the 'official version' the artistic model of the Arvals' calendar; and only indeterminate traces of religion.

The stone calendar is not a witness in stone to the religion of ancient Rome. But may it lead us to such a witness? The conclusions we have arrived at so far do not exclude the possibility of an official version (on the Forum? before the official residence of the *pontifex maximus*?).[44] They may well make its existence seem improbable for the Augustan Age, the period of the extant, defective *fasti*. But the model for the calendar must have its origin somewhere; it must, somehow, have been already present in the Republican Period. It seems unlikely that the credit for it should be given to the painter of the Fasti Antiates maiores, to whom we owe the sole extant Republican specimen. Clarification of this 'somewhere' and 'somehow' is sought in the following section. The quest entails a reconstruction of the history of the Roman calendar. It does not confine itself to the calendar's technical development, but focuses rather on the question of the 'official version'. It is thus concerned with giving a history of the form of the Roman calendar, and of its foundation in reality.

[43] SCHEID 1992: 121.

[44] The fragments of the consular *fasti* were displayed in front of the Regia until the 1940s (see TAYLOR AND HOLLAND 1953: 137)! SCHÖN 1909: 2034ff. argued against the *pontifex maximus*.

3

Towards an Early History of the Roman Calendar

3.1 Notions of a Prehistoric Calendar

As I remarked at the beginning, any reconstruction of the earliest Roman calendar system must begin with Late-Republican and Imperial-Period accounts and theories regarding its development. A radical critical approach is needed in order to obtain such an understanding; this is demonstrated in the case of the theory, propounded in ancient texts, of a pre-Republican, 'Romulan' ten-month year. The evidence is circumstantial and inadequate. All speculations on the subject depend on the numeric month-names, which had evidently been reckoned from March, and not from January, when the calendar year currently began. The resulting, strictly *a priori* interpretation, that, historically, the year had had a different starting point, is demonstrably false. Relatively speaking, different 'years' within an integral calendar context (with its own 'beginning') are more the rule than the exception. When ancient authors reckon the Romulan year as comprising 304 days, this corresponds to the sum of the first ten months of the year (January to October), on the basis of month-lengths in the Julian-Augustan calendar, showing that the ten-month year was a late construction. All that can be said for certain is that the Republican calendar was preceded by a calendar with twelve genuinely lunar months, six of which incorporated the names of gods, while six bore numeric names.[1]

The Roman antiquarian Varro refers to a law of the early fifth century that, he says, used a dating system with *mense intercalari*;[2] this demonstrates that, rather

[1] BRIND'AMOUR 1983: 225.

[2] Macrob. *Sat.* 1.13.21.

The Roman Calendar from Numa to Constantine: Time, History and the Fasti, First Edition. Jörg Rüpke.

than a pure lunar year, the pre-Republican calendar followed a lunisolar system: a year of twelve lunar months, which, rather than constantly shifting against the solar year and the seasons, like the Islamic year, was continually corrected by the intercalation of a thirteenth month, as in the Jewish calendar. Until the Late Republic, and perhaps later, the key days of the month were associated with religious institutions that can have originated only at a time when the calendar comprised real lunar months. These days can readily be linked with phases of the moon.[3]

3.2 The Structure of the Month

The antiquarian tradition is unanimous in seeing the *idus* as representing the old days of the full moon.[4] On these days, the *flamen Dialis*, attended by other *sacerdotes*, sacrificed a white ram *(ovis idulis)* to Jupiter.[5] This lavish ritual, the association with Jupiter, and the *NP* indication in the later *fasti*, indicate that the 'Ides' month-divider was the most important point of orientation in the month.

If the Ides mark the full moon and the middle of the month, then a correlation must be established between the Kalends and the new moon. We may assume that specialists in Rome also saw the necessity for such a correlation. Although the etymologically orientated accounts in respect of *kalendae* do not offer the conjecture, it is mentioned by Macrobius for one.[6] His formulation turns out to be very precise: *exortu lunae* does not refer to our new moon ('dark moon'), but to the first appearance of the waxing moon, the 'young crescent' (the 'new moon' properly speaking).[7]

The etymology encouraged the depiction of a bizarre ritual. A *pontifex minor* – a pontifical scribe only later referred to by this title[8] – observes the moon. He passes on the result of his observations to the *rex sacrorum*. The object is to determine the number of days until the observance of the Nones. Both priests enter the Curia Calabra, where they make a sacrifice. Their tasks then diverge. The pontifical aid invokes Juno Covella, repeating the formula *kalo Iuno Covella* five or seven times, to indicate the period that must elapse before the Nones.[9] The assembled people are then informed of the result in similar or identical terms: the

[3] Dion. Hal. 16.3.2.

[4] Macrob. *Sat.* 1.15.14–16; Lydus, *Mens.* 3.10 (p. 47 Wünsch) and Plut. *Quaest. Rom.* 24. An alternative account in Suetonius: Isid. *Nat.* 4 (= Suet. *frg.* p. 166.4–7).

[5] Ov. *Fast.* 1.56; Festus, *Gloss. Lat.* 372.8–12 L; Paul. Fest. 93.3 L; Macrob. *Sat.* 1.15.16; Lydus, *Mens.* 3.10 (p. 47 Wünsch); cf. Varro, *Ling.* 5.47.

[6] Macrob. *Sat.* 1.15.19–20.

[7] Macrob. *Sat.* 1.15.5; Plut. *Quaest. Rom.* 24.

[8] See Liv. 22.57.3–4.

[9] Varro, *Ling.* 6.27; Macrob. *Sat.* 1.15.10.

sources do not specify which.[10] Perhaps at the same time as this is going on, the *regina sacrorum*, wife of the *rex sacrorum*, sacrifices a pig or a sheep to Juno in the Regia.[11]

The facts appear sufficiently clear. No further surmises can be built on the etymologies, all of which prove to be unreliable. The association between Juno *Covella* and the fundamental term *kalo* is impossible to reconstruct;[12] but that between *kalo* and *calare/calata* is evident: *calare* relates to the convening of any kind of people's assembly; ancient conceptions have this as the derivation of *classis*,[13] and of the name of the Curia *Calabra*.[14] These associations cannot, however, be confirmed linguistically. Although the particular epithets concerned, of the goddess, *Covella*, and the location, *Calabra*, therefore remain opaque, the ritual itself is enlightening as to the nature of the relationship between the time of month and the deity: the invocation of Juno Covella for each day leading up to the Nones may be understood as part of an epiphany ritual that does not conclude until the Nones, when the period in question passes into the ordered structure of the month, having been uncertain until then owing to its variable length. A conceptual link between Juno and the moon is, therefore, undeniable.

The direct function of the Kalends lay in preparing for the Nones. Ancient interpretations speak of associations with the moon's phases, but vary in their terminology. The Augustan, Greek antiquarian Dionysius speaks of the first quarter,[15] and there is the link in popular etymology between *nonae* and *nova luna*, although this can no longer be equated terminologically with *principium lunae*.[16] Only the self-evident association between *nonae* and *a. d. nonum idus* is acceptable in linguistic and practical terms.[17]

As to the function of the ritual, the ancient authors are in agreement: the *rex sacrorum* announced the feast days *(feriae)* to be celebrated during the month. We have no way of knowing how comprehensive his proclamations were. Varro (116-29 BCE), describing the proceedings as an eyewitness, shows some reticence: by his account, they may have confined themselves to the names of the feasts.[18] Macrobius, on the other hand, writing four hundred and fifty years later in the early fifth century CE, and in the past tense, already seeks a sociological context for the ritual. We are unable to determine the extent to which 'reasons' *(causas)* and rituals or instructions for the ritual *(quod ... faciendum esset)* reflect information beyond that to be found in Varro, or simply serve as decoration. It is permissible

[10] Verrius Flaccus, *Inscr. It.* 13.2.111 (*fast. Praen.* 1 Jan.); Serv. *Aen.* 8.654 (*auct.*); Macrob. *Sat.* 1.15.9–12; Lydus, *Mens.* 3.10.

[11] Macrob. *Sat.* 1.15.19.

[12] WALDE AND HOFMANN 1.282; ERNOUT AND MEILLET 146.

[13] Macrob. *Sat.* 1.15.11; Quint. *Inst.* 1.6.33.

[14] Varro, *Ling.* 5.13. An alternative account in Serv. *Aen.* 8.654. See also Paul. Fest. 42.16–17 L.

[15] Dion. Hal. 16.3.2; Lydus, *Mens.* 3.10.

[16] Varro, *Ling.* 6.28; Festus, *Gloss. Lat.* 176.20–24 L; Macrob. *Sat.* 1.15.12–13. Cf. Plut. *Quaest. Rom.* 24.

[17] Cf. Isid. *Orig.* 5.33.14.

[18] See also Varro, *Ling.* 6.13.

to read the brief instructions as to the festivals to be celebrated, and the behaviour to observe for the cult ('Come wearing a wreath'; 'Bring arable produce with you'; etc.), as contents of the proclamation. There is no indication that the *rex sacrorum* recited cult myths at this point. The wording he used is not referred to by any antiquarian or mythographer.

We know nothing of any further cult actions performed on the day, apart from occasional temple-dedication feasts on the Nones. The *sacra nonalia* spoken of by Varro comprise just those proceedings that he describes; they do not represent any autonomous event.[19] Overall, then, this was a day devoted above all to practical purposes.

One further date may be added to the monthly structure outlined so far. It can be demonstrated that, mirroring the Nones, which fell on the ninth day – always counting inclusively – before the Ides, there was a monthly date on the ninth day after the Ides. The idea of a missing, fourth 'orientation day' was first developed by THEODOR MOMMSEN.[20] Although, however, he rightly took as his point of departure the Roman nundinal week, he mistakenly derived it from the phases of the moon.[21] Beginning from the definition of the *nundinae* as *dies fasti*, he accordingly looked for days marked with an *F* eight days before the Kalends.[22] But this way of proceeding forces so many assumptions, and results in such a complex system of irregular weeks, that it encountered justified criticism.[23] The existence of such a fourth date, is, however, required by considerations that follow regarding the history of the *nundinae*. Thus the evidence provided by indications in inscriptions three hundred years more recent does not impinge centrally on our reconstruction of the monthly structure, although the inscriptions themselves do have something to tell us about the history of the calendar's written form, and details of the month's ritual structure, and so merit at least brief consideration.

In the *fasti*, the traces of this monthly cult structure are concealed in the Tubilustrium, which fell, with the *QRCF* day, on the 23rd and 24th of March and May. Two converging lines of argument lead to this conclusion. We must first turn our attention to the second component of the double date, the day marked *QRCF*. MOMMSEN himself remarked on the striking duplication of the 'double feast' *Tubilustrium / QRCF* (23 / 24 March and May). The abbreviation (*Q. Rex C. F.* in the Fasti Viae Lanza) can be resolved as *Quando Rex Comitiavit Fas.*[24] Neither the cult proceedings involved nor the meaning of *comitiare* is obvious. There are uncertain

[19] Ov. *Fast.* 1.57: *Nonarum tutela deo caret.*

[20] MOMMSEN 1858: 215–40.

[21] MOMMSEN 1858: 215–16.

[22] MOMMSEN 1858: 226.

[23] See MOMMSEN 1858: 231–2, 236. Crit.: FLEX 1880: 11–13.

[24] Varro, *Ling.* 6.31: *Dies qui vocatur sic 'Quando rex comitiavit fas', is dictus ab eo quod eo die rex sacrificolus dicat ad Comitium, ad quod tempus est nefas, ab eo fas: itaque post id tempus lege actum saepe;* Paul. Fest. 311.1–3 L (cf. Festus, *Gloss. Lat.* 346.22–36 L).

indications of a morning sacrifice.[25] The interpretation 'to hold a committee'[26] finds little support in the sources, which envisage the location *Comitium*; on the other hand, it must count as a good solution from a technical point of view. MOMMSEN made it the basis for his argument linking the two *QRCF* days with the *comitia calata*,[27] which were held only twice annually under the leadership of a priest.[28] Despite lingering problems, the interpretation gives us the key to our solution: the similarity of the *comitia calata*, with its gatherings for promulgating information, to the festive programme on the *nonae*. The points of similarity lie in the protagonist, the *rex sacrorum*, in the non-decision-making character of the assemblies (the *comitia calata* merely involved a higher degree of formality), and in terminology: in both instances, the term *kalare*[29] played a central role.

Closer analysis shows that the correlations go much further. The *QRCF* days each form the close of a two-and-a-half day *nefas* period (from the 22nd to the 24th), at whose midpoint is the Tubilustrium. The Tubilustrium also comes one *nundinum* after the Ides, as the Nones comes one *nundinum* before. The end of the incomplete passage from the Imperial lexicographer Festus[30] confirms what we know from elsewhere about the *comitia calata*:[31] that a similar relationship and interchangeability between the *rex sacrorum* and the *pontifex maximus* pertained on the *QRCF* day as on the Kalends and the Nones. Just as a lamb was sacrificed in the context of the Tubilustrium,[32] so – separated by the '*kalatio*' – on the Kalends, and, what is more, to Juno by the *regina sacrorum*, wife of the *rex sacrorum*.[33] It might also be possible, even without the conjectured but probably to be rejected *descendat* in the Varro passage,[34] to identify the choreography of the *QRCF* assembly with that of the Kalends (Arx, Curia Calabra); here, we should imagine the *sacrificium* as taking place on the Arx. The actual assembly then took place on the Comitium: the same location as may have been used on both Kalends and Nones, and the site of the Regifugium.[35]

25 See Varro, *Ling.* 6.31; Festus, *Gloss. Lat.* 346.22–36 L: *sacrificium*. Probably *sacrificium in honorem* in Festus, *Gloss. Lat.* 310.12–21 L: *[Q. R. C. F. Quandoc Rex co]mitia sit (sic* Mss.*) [fas ...]ṇ honorem [...] menstruis scrip[...] rege dicuntur. [... scriptori]bus traditae sunt. [...]r pars ante[rior ...] posterior [...] si quis alius pro rege [... pon]tifex, tum is dies [...].* The sacrifice is confirmed for the structurally similar Regifugium: Festus, *Gloss. Lat.* 346.27 L; Plut. *Quaest. Rom.* 63.

26 See MAGDELAIN 1980: 6.

27 Gai. *Inst.* 2.101–3. MOMMSEN 1858: 229.

28 Lael. Fel. *Gell.* 15.27.1–3.

29 MICHELS 1967: 37; MAGDELAIN 1980: 10.

30 Festus, *Gloss. Lat.* 310.19–20 L.

31 KÜBLER 1897; SZEMLER 1978: 360–3; WOLF 1980.

32 Festus, *Gloss. Lat.* 480.25–9 L: *[Tubilustria,] quibus diebus adscribtum in [Fastis est, in atr]io Sutorio agna tubae [lustrantur, quos] tubos appellant; quod genus [tubae Euander ex Ar]cadia* Pallanteo trans[tulisse Lutatius scripsit*]* (*suppl. RÜPKE; *lustrationis ex Ar – latum esse dicunt* suppl. URSINUS; cf. Verrius Flaccus, *fast. Praen.*; on Arcadia cf. Liv. 1.51).

33 Macrob. *Sat.* 1.15.19. ROSENBERG 1914, 1914a.

34 Varro, *Ling.* 6.31.

35 Plut. *Quaest. Rom.* 63.

That this fourth date belonged in the old nundinal structure is supported, first, by Festus' reference to Etruria,[36] and, second, by the common, promulgating function of all the days mentioned here (particularly important at the *comitia calata* in respect of wills).[37] The strongest argument is provided by the word *menstruis* in the damaged Festus lemma '*QRCF*'.[38] It necessarily indicates an earlier monthly occurrence, and came down to Festus through at least one author, whom he quotes: *scrip[...]* ('wrote') can still be read in the Codex Farnesianus.[39] The relevant sources are unable to tell us why the *QRCF* came to be celebrated only twice annually; an answer to this question should be sought in the first component of the double date, to which we must now turn.

We know some ritual details of the Tubilustrium on 23 March. The 'purification' of the *tubae* took place in the *atrium sutorium*, the 'house of the shoemakers',[40] whose location is unknown. Instrumental in the ceremony was a female sheep *(agna)*. The objects to be purified were *tubi,* a kind of *tuba* used in sacral contexts.[41] Responsible for carrying out the ritual appear to have been the *tubicines sacrorum* themselves,[42] evidently cult personnel standing at the threshold of the priesthood.[43] All these details apply similarly to 23 May.

The syntax of the ritual provides no confirmation in itself of the interpretation of 'purification', presented in the sources. Apart from their annunciatory function at the games, the main use of *tubae* or *tubi* was at funerals. There is one other instance, however, and the only one that brings us further forward: not known to us, moreover, from routine rituals, but only from discrete events, is their use at eclipses of the moon. The witnesses created a cacophony. They used all possible means, but *tubae* in particular, to come to the aid of the moon as its light was extinguished.[44]

This gives us an interpretation for the Tubilustrium that fits much better with the dates: the *tubilustria* were monthly rites, one *nundinum* after the Ides full moon. Their function was to fortify the waning moon (Luna). One expedient was the sacrifice of a white sheep; but the main means used was to make a noise with trumpets, as we find variously and widely attested in respect of eclipses of the moon.

If the Tubilustrium was at one time a monthly occasion, why did it largely disappear from the calendar, persisting only in March and May? The question itself needs to be stated more precisely. We cannot exclude the possibility that the rudiments of the cult persisted on the 'lost' dates: the calendar, however, used them

36 Festus, *Gloss. Lat.* 346.33 L.

37 MAGDELAIN 1980: 7.

38 Festus, *Gloss. Lat.* 310.12–21 L.

39 Festus, *Gloss. Lat.* 310.14 L.

40 Varro, *Ling.* 6.14; *Inscr. It.* 13.2.123 (*fast. Praen.* 23 March); Festus, *Gloss. Lat.* 480.25–9 L.

41 *Inscr. It.* 13.2.123 (*fast. Praen.* 23 March); Varro, *Ling.* 6.14; 5.117. Cf. Calp. *Ecl.* 1.67–8.

42 Festus, *Gloss. Lat.* 482.27–9 L.

43 RÜPKE 2008.

44 Juv. 6.442–3; Tac. *Ann.* 1.28.1–2; Liv. 26.5.9.

eponymously only on these two dates. But do the two days really owe their presence in the calendar to the lunar ritual? If, normally, a day's name forms the basis of its character as *NP*, which in turn indicates *feriae*,[45] it is striking that the date in March is referred to as *feriae Marti*, and the one in May as *feriae Volcano*.[46] It is not unusual to find *feriae* on the day of the month that would mark the Tubilustrium. Other festivals, even important ones, fall on such days: the Terminalia in February,[47] the Parilia, 'birthday of the city', in April; the Neptunalia, a kind of Roman Feast of the Tabernacles,[48] in July; the Consualia in August, and finally the 'Divalia' in December. In months that have *feriae* in their final third, there is always a festival on the day one *nundinum* after the Ides. A political analysis of the orientation days shows this to be unsurprising: the pre-decemviral assembly day falls precisely opportune.[49]

The consequence for the Tubilustrium was that this routine monthly ritual was so obscured by a succession of popular celebrations, that it could not establish any exalted identity of its own. A further reason for the absence of *feriae* status was the lack of a clearly formulated association with a deity, such as was provided by Jupiter on the Ides, and, in a limited sense, Juno on the Kalends. In other words: the ritual in the *atrium sutorium* took place on a monthly basis (*menstruis*; probably excepting February), at least long enough for it to leave traces in the antiquarian tradition; but, in the normal run of events, it ceased to excite further notice.[50] It is also worth remarking that, in the *tubicines*, it was served by cult personnel of only secondary standing. In the light of the day's status in the calendar, and the general demise of lunar awareness in Roman religion during the course of the Republic, the *pontifices* and *tubicines* finally limited its observance to two dates, while the *rex sacrorum* and the *pontifex maximus* could act similarly in respect of the *QRCF* on the following day, owing to its reduced practical significance.

To return to the entries in the calendar: in March and May we have the situation that both *feriae* are connected with deities that have other feast days elsewhere. Vulcanus has the Vulcanalia on 23 August, while Mars has *feriae* on 1 March, perhaps on the March Equirria (thus probably in February too), and, according to the Fasti Vaticani, also on the Quinquatrus.[51] In these circumstances, it would scarcely have seemed apt to register another Vulcanalia, even less a Martialia; even if, at first glance, the Consualia appear to show that this would not have been impossible. A further factor is that the cult profile of concurrent festivals on the two remaining Tubilustrium days was unusually meagre: only a *sacrificium*

45 See Ch. 5.2.
46 March: *Inscr. It.* 13.2.104. 123; May: 13.2.57. 187.
47 See Ch. 6.2: to be taken into account here is the two-day delay at the year's end, assured by the Regifugium.
48 See Paul. Fest. 519.1–2 L.
49 See Pl. *Leg.* 738cd.
50 An alternative view in HUMM 2005: 473.
51 *Inscr. It.* 13.2.121 (Fast. Praen. 1 March); 173 (Fast. Vat. 14 and 19 March).

is mentioned for 23 March.[52] But it was precisely this that provided the grounds for making the routine ritual the eponymous factor in the calendar published by Cn. Flavius at the end of the fourth century.

A brief summary of our conclusions so far: the lunar month was organized on the basis of observation of the 'new moon' at the Kalends; on this day was announced the date for the Nones, and on that day in turn the month's festivals were announced to a further assembly. There followed, in a fixed, nine-day cycle (by Roman reckoning), thus at nundinal intervals, the Ides (full moon), and, at the waning moon, the so-called Tubilustrium. The particular chronological relationship between Kalends and Nones had a clear function: a period of variable length governed the synchronization of the monthly social cycle with its counterpart in nature. After the adjustment at the Kalends, no further such measure was required for the rest of the month, as neither the Nones nor the Tubilustrium demanded an absolutely precise lunar phase. The entire structure corresponds to the calendar postulated by KARL OTFRIED MÜLLER for the Etruscans on the basis of the same sources, and that same calendar must also have applied in Rome into the Early Republic.[53]

Of particular import here is the *feriale* referred to as the Tabula Capuana, a ritual calendar from the period *c.*470 BCE. This text governed the cult at the Haema sanctuary, which was situated near Capua, and must have been a regional cult centre.[54] In contrast to the much later Etruscan cult calendar in the form of a text written on linen, which came to light in the guise of the 'Agram mummy-wrappings', the tablet from Capua, now preserved in Berlin, does not number the days of the month. For each month, only a few, recurring days are specifically named. The first such day listed is usually the Ides, thought by the Romans to be of Etruscan origin,[55] and here called *iśveita*. If we can give credence to the Late Antique informant Macrobius and his Late Republican sources, the Etruscans had a periodical system that had 'Nones' repeated at eight-day intervals. In that case, they may have had 'weeks', based on a series of four structural days (Nones–Idus–'Tubilustrium'–Kalends) beginning anew each month, just like the Roman calendar described above. The dates shown on the Capuan tablet fit precisely with this structure: the Ides, marking the beginning of each month, was followed, probably at intervals of a week, by the days *celuta* (waning moon, the Roman 'Tubilustrium'), then *tiniana*, associated with the sky deity Tinia (the Roman Kalends, dedicated in Rome to the goddess Juno), and finally *aperta*, the Roman Nones, at the waxing moon. The only day-name comparably formed on the basis of a number in this Etruscan text, *macvitule*, incorporates the numeral five, and (assuming Roman inclusive reckoning) would thus fall in the middle of the 'week' formed as described. Owing to the few sources available for the period in question, we cannot know

52 Ov. *Fast.* 3.850.

53 MÜLLER 1828: 2.323–8.

54 CRISTOFANI 1995. On what follows, RÜPKE: review in *Gnomon 1999*, 272–4.

55 Macrob. *Sat.* 1.15.14; weeks: *ibid.*, 13.

who adopted whose calendar here. It is enough to have established that the city of Rome and its calendar can be seen as a variant of Middle Italic practice in respect of calendars.

This reconstruction also clarifies the function of the Kalends. This cannot have been the empirically determined day of the new moon: how could the people attend an assembly whose date they did not yet know?[56] There is an additional practical difficulty. It is difficult to imagine that direct observation of the new moon could have been used for the calendar. The moon's crescent may become visible on the first, the second, or even the third day after the new moon as determined astronomically.[57]

These difficulties are resolved by the fixed period between the Tubilustrium and the Kalends. It enabled people to know when they had to attend in the city, or more precisely on the Arx, to hear the crucial announcement. It also made the second half of the month considerably longer than the first, so that the Kalends did not fall at the precise time of the moon's invisibility, but at about the time the first, narrow lunar crescent appeared. This made the system's operation much simpler. The pontifical aid could estimate the period to the next full moon by observing the size of the lunar crescent, and set the date of the Nones accordingly, so that, nine days (inclusive reckoning) later, the Ides would coincide with the full moon.[58]

The system gave the position of the Kalends a certain ambiguity: on the one hand, they were firmly bound into the structure of the old month; on the other, they opened the new. It must not be forgotten that, although the new month ('moon') had visibly begun, the month's ritual programme, and originally also its political programme, had still to be announced on the Nones, thus at the far side of the variable period linking the months. It is in the day's ambivalence that we must seek the explanation as to why the period between the 'ninth day after the Ides' and the next Kalends comprises not nine, but ten days in the Roman reckoning. This may be a consequence of the calendar's committal to written form, and the necessity of resolving the day's ambiguous status at the month's limit by assigning it firmly to one month or the other: in the event, to the one already begun. The day now missing from the previous month was inserted immediately before the beginning of the current month, and thus the Kalends in a manner of speaking duplicated. It may have been this circumstance that motivated Caesar to add his new days and make his additional intercalation *before* the last day,[59] and thus secure the connection between the old month's ritual ending, the detail of which is unknown to us, and the next Kalends. The two-day structure of the

[56] Cf. the Jewish practice of announcing the *calculated* date of the new moon (*molad*) during the Sabbath morning service preceding the week of the new moon (A. Spier 1952: 217).

[57] Cf. for other cultures Parker 1950: 10; Derchain 1962: 30; Zion Wacholder and Weisberg 1971; Lasker, Lasker 1989; Ilyas 1982; Basnizki 1938: 12–16; Strobel 1977: 71–3; Robbins 1989: 136.

[58] This observation too can already be found in Müller (1828: 2327).

[59] See Ch. 8, n. 29.

Tubilustrium/*QRCF* complex, with its surely very ancient distribution of the festival and assembly functions over two days, had provided a precedent that, so far as the priesthood was concerned, already disrupted the nundinal structure.

Schema of the Pre-Republican Calendar with Empirically Determined Lunar Months

Kalendae = first moonlight/narrowest crescent

|

(variable: 3–6 days)

|

Nonae

|

(7 days)

|

Idus = full moon

|

(7 days)

|

'Nundinae post Idus'/*'Tubilustrium'*

|

(7 days)

|

Kalendae = first moonlight/narrowest crescent

|

(variable: 3–6 days)

|

Nonae

|

=

=

3.3 Market Cycles

What was the status of the 'orientation days', now increased to four? The little that the sources have to say on the subject includes festival elements as well as characteristics of market days. The latter might be said to predominate in the case of the Nones, when the rural population was accustomed to crowd into the city. The case of the Kalends is similar. The Tubilustrium/*QRCF* complex likewise shows signs of being predicated on the necessity for an assembly. The Ides bring cult events, the festal element, more strongly to the fore; although, here too, the established 'weekly cycle' also suggests market functions. In terms of ethnological comparison, what we see here is an entirely normal model for agrarian societies.[60] The most familiar parallel is the *šabattu,* later to become the Jewish Sabbath, whose original function was as a day of assembly and feast day at the full and new moons.[61]

In terms of both functional description and rhythm, the orientation days coincide closely with the 'actual' market days known as *nundinae*. In the Republican and Imperial-Period calendars, the *nundinae* form a constant, weekly, eight-day cyclic pattern independent of the calendar.[62] While the main function of this weekly cycle was to govern the economic relationship between the city and the surrounding countryside by means of its market day,[63] it also appears to have created the effect of a 'Sunday break'.[64] The day's character is ambivalent in terms of its intended purpose: on the one hand, it counts as *feriae*; on the other, the law declares it *dies fastus*, a possible court day, thus taking account of the legal needs of the rural population.[65] In facilitating public announcements, the *nundinae* formed a fundamental basis for juridical and political diary dates.[66] Cult activities attached to the *nundinae* themselves were constrained in nature, and not public. The *flaminica* sacrificed a sheep to Jupiter in the Regia.[67] In its form, the nundinal cult may be regarded as a synthesis of elements of the cult observed on orientation days.

Taken together, the points of convergence between the social and cult contexts make it appear extremely improbable that the two systems continued to operate in tandem within one, still functioning lunar calendar. Ancient as the nundinal cycle as such may have been, the system of *nundinae* may be best understood if it is seen as replacing a system in which the orientation days also served as market days.[68] Such a system of fixed, 'natural' market days, associated, on the Greek model, with particular days of the month, still existed in the Eastern Roman

[60] WEBSTER 1916: 103–17.

[61] See LANGDON 1935: 90–7.

[62] LINTOTT 1968: 189.

[63] See Rutilius Rufus in Macrob. *Sat.* 1.16.34; Varro, *Rust.* 2, pr., 1; Festus, *Gloss. Lat.* 176.24–7 L; Isid. *Orig.* 5.33.14.

[64] Varro, *Sat. Men.* 279.

[65] WISSOWA 1912: 440 n. 2. Gran. Lic. Macrob. *Sat.* 1.16.30. See Hemina, *Hist.* 14 Peter = Macrob. *Sat.* 1.16.33; Rutilius, *Hist.* 1 Peter = Macrob. *Sat.* 1.16.34; Plin. *HN* 18.13.

[66] LINTOTT 1965. For a full account of the *trinundinum* see BRIND'AMOUR 1983: 87–96.

[67] Gran. Lic. Macrob. *Sat.* 1.16.30.

[68] Thus also MÜLLER 1828: 2.325–6.

Empire in the Imperial Period:[69] the nundinal structure of the orientation days, with fixed intervals between Nones, Ides, 'Tubilustrium', and Kalends, having by this time been disrupted by the calendar reform, and lost many of its functions.

We even have testimony for this 'late' origin of the *nundinae*, even though the accompanying justification must be rejected. While our sources for the decemviral intercalation law, the second-century BCE historians Cassius Hemina and Sempronius Tuditanus, ascribe the establishment of the *nundinae* to Servius Tullius, or even Romulus (who appears here along with Titus Tatius as a founder of religious institutions), Tanusius Geminus and Varro assume a Republican origin.[70] The creation of the *nundinae* may be regarded as the central component of the calendar reform, and cannot be separated from the establishment of the nundinal letters: it was only through these that a written calendar became practicable.

3.4 Modes of Dating

In their character as market days, the month's four structural orientation days were effectively equal in status. As far as their technical status in the calendar is concerned, however, differences must be noted. The most important of these is our inability to assign a name with any certainty to the fourth monthly market day concealed behind the 'Tubilustrium' label. The simple numeric form embodied in *nonae* shows us that we should not expect such a name to be a striking one. A possibility would be to call this day 'the ninth day after the Ides': forward reckoning is sufficiently attested by the feast-day title *Quinquatrus*, five days (again, by Roman inclusive reckoning) after the Ides, and the formulations *nonae quintanae* and *septimanae* defining the interval forwards from the Kalends.[71] If the straightforward solution *nonae post idus* is felt to be unacceptable (although the Etruscans are indeed credited with more than one instance of *nonae* in the month),[72] *nundinae*[73] or **nonatrus* (unattested, but a possible coining as similar in meaning) might be conceivable.

The problem of the name (and thus, as has been demonstrated, that of identification) is exacerbated by the frequent accumulation of significant festivals on the *nonae post idus*. The evocative feast-name (Parilia, Neptunalia) may have tended to obscure the day's run-of-the-mill generic name (*nonatrus*?). If observations con-

69 Nollé 1982: 45; Ligt 1993; Andreau 2000.

70 Cassius Hemina (*Hist.* 14), Tanusius Geminus (*Hist.* 4 Peter, second half of the second century BCE, and Varro in Macrob. *Sat.* 1.16.32–3.

71 For the Nones: Varro, *Ling.* 6.27; Macrob. *Sat.* 1.15.10; Gagnér 1922: 202–11. *Quinquatrus*: Varro, *Ling.* 6.14; Festus, *Gloss. Lat.* 304.33–306.8 L.

72 Macrob. *Sat.* 1.15.13.

73 Mommsen 1858: 226.

cerning traces of old dating practices prove to be correct, the Kalends and Ides alone originally served as points of reference for a forward and backward reckoning extending over a maximum of nine days. In this event, there would not even have been any need to use the generic name of the fourth orientation day in dates. When, with the reform, the interval between Kalends and Nones became fixed, there may have been a transition – perhaps only gradual – to consistent backward reckoning, which would now also have been practicable in respect of the Nones. This backward reckoning then, by analogy, displaced forward reckoning from the Ides, and was implemented from the Kalends back through to the Ides, even though the numbers became quite high in consequence. The first date with an interval-number of nine or more (i.e. for the 'week' after the Ides) does not appear in the Fasti triumphales Capitolini until the 46th entry, in 343 BCE; with twenty instances in the following 168 entries, this shift is statistically significant at a level of 1 per cent.

Like other Mediterranean-world calendars operating with lunar months, the Roman calendar too endeavoured to establish a stable relationship with the solar year, which determined the course of the natural year, agriculture and seafaring, and, last but not least, opportunities for warfare. This reconciliation was encompassed in the calendars we know of by the occasional intercalation of a thirteenth month, whether under a specific name or derived by merely duplicating one of the twelve normal months. The inclusion of such a month every two or three years kept the relationship between lunar calendar and solar year constant over the long term. Such systems must be referred to as lunisolar rather than lunar calendars. They more or less succeed in reproducing both the periodicity of the Earth's orbiting satellite and the parent planet's path around the central star.

Easily applicable formulas can be derived mathematically for implementing such systems. As early as the sixth century BCE, the Greeks knew that including such a thirteenth month three times every eight years – the so-called *octaeteris* – produced a very precise result. It was the remaining excess of some one and a half days in this cycle that moved Meton and his pupil Euctemon, probably in 432, to devise one that was still more precise. A total of 235 months, 125 of them comprising thirty days, 110 twenty-nine days, was to be accommodated in a nineteen-year cycle.[74] This 'Metonic cycle' evidently relied on an existing conventional lunar month, observation of the moon being replaced by the consistent alternation of months with thirty and twenty-nine days, to correspond to a lunation averaging 29.54 days. Devices were even fashioned for its implementation: the 'Anticythera mechanism', found in a shipwreck off the island of Anticythera in 1900, used a differential drive and more than thirty cogwheels to synchronize the lunar year, labelled with Greek month-names (outer ring), with the solar year. In this way, it permitted its operator to track the sun's position in the Zodiac (inner disc), while

[74] Briefly VOGTHERR 2001: 30–1; a full account in SAMUEL 1972: 42–9. On Callippus' improved 76-year cycle see recently A. JONES 2000.

also translating it into 'civil' dates. The mechanism's action is predicated on the Metonic cycle.[75]

All actual historical dates we know of suggest that these cycles never entered into regular use in the important Greek cities. Rulers frequently succumbed to the temptation of lengthening their own period in office or lease on authority, or increasing their power of disposal, by intervening in the system's 'clockwork': the fatal consequences can be particularly well observed in the case of Rome.[76]

Such adjustments were also made in Rome, whose calendar had its own aspirations to become lunisolar. We have the already quoted accounts of attempts to regulate the intercalation by law, dating from as early as the fifth century.[77] This only shows how controversial the question was. Not only did the first Roman calendar revolution – we can hardly call it by any other name – prove unable to bring about change in this area; probably, in order for the project to succeed, the very attempt had to be foregone. It is no accident that the problem was not resolved until the dictator Caesar himself drew a line under it.

For the fifth century, the Roman calendar may be regarded as being fully integrated with its counterparts in the regions and across the territory of Italy. The phases of the moon were everywhere the same; although what the particular months were called, and when and how intercalations were effected, may have varied from place to place. Every fifteen kilometres a new name for the month was still the rule rather than the exception in the parochial political units of Italy and Greece. This was all to change for Rome.

In the course of the fourth century, Rome had turned from its original condition, as a city dominated by Etruria, into the major power of Central Italy. Militarily, this development was brought to a conclusion by the victory in the Latin Wars, which was achieved between 340 and 338. Rome began to set up colonies, and to consolidate its hegemony by occupying new lands: from now on, mobility was to become a characteristic of Roman citizens. The precondition for military success was an internal political process that has been identified as the emergence of the Roman nobility. A 'historic compromise' redirected the embittered rivalries of the old noble families, the patricians, both among themselves and with the leading representatives of the plebeians, who had gradually acquired more rights and positions. Power struggles were diverted into activities that benefitted all, especially in terms of military success. It was no longer personal booty that should count, but victory as an elected magistrate (consul, *praetor*) and commander. Only such positions provided the justification for embarking on 'international' conflicts; but these magistrates did not have only their own slaves, clients and family members to draw upon for troops, but the Roman army, constantly enlarged by the forces of allies. The 'common cause', the *res publica*, was, as it were, remoulded,

75 Ill. in RÜPKE 2006b: 25.

76 For Greece: PRITCHETT 1999; on early criticism: W. R. CONNOR 1987.

77 Macrob. *Sat.* 1.13.21.

and organized in new institutions. At the same time, the code of honour became focussed on these institutions and political processes; the bywords 'honour', 'courage', and 'concord' redirected egoisms and rivalries towards socially useful causes, and were revered as divinities in newly built temples (paid for with the booty of war!).[78]

To dramatize is always perilous. Rome's rise from peasant village to world power extended over more than half a millennium, and was accompanied by many setbacks. Her supremacy is thus more easily apparent in hindsight. The victories of the third, and even in part the second century, belonged to the Latins and the Italic peoples. Even in the late second century BCE, cities such as Praeneste still sought to rival Rome in the creation of monuments and prestige buildings; Rome's success in carrying off the victory in the war against her own allies during the years 91 to 89 could not be foreseen at the time. Although Rome's power was supreme, she had competitors, some of whom were only a little weaker.

78 Seminally HÖLKESKAMP 1987; RÜPKE 1990; FLAIG 2003.

4

The Introduction of the Republican Calendar

4.1 Timing and Motivation

The most important step in the Republican history of the calendar was the transition from months corresponding to the phases of the terrestrial moon to conventionalized months. These were harmonized with the solar year by means of regular intercalations, but, as they no longer corresponded to lunations, were no longer lunar months.

Roman tradition usually associated this transition with Numa,[1] the great legislator, the king supposed to have laid the religious foundations of the state.[2] The grounds for this view may possibly be found in the peculiar number-relations characteristic of the 'new' months. Month-lengths of twenty-nine and thirty days, which would come closest to the average 29.53 days of a synodic month, and normally characterize the structure of 'conventional' lunar months, do not occur.[3] Instead, we see months of twenty-nine and thirty-one days. The year, with its 355 days, is one day longer than a year comprising twelve synodic months. *Impar numerus*, the odd number, was regarded as propitious, and the sources quoted show that the numerology was seen as being Pythagorean. In 181 BCE, writings of Numa 'discovered' in a stone chest clearly showed him to be a 'philosopher'.[4] The characterization 'Pythagorean', to be more accurate Pythagorean *ante litteram*, became attached to the attribution.[5] Hence, with a logic that, although not

1 Outlined in Liv. 1.19–21, with OGILVIE 1965: 95–6.

2 See Cic. *Leg.* 2.29; Liv. 1.19.6–7; Plut. *Numa* 18.1–3; Macrob. *Sat.* 1.13.1–5; cf. Censorinus, *DN* 20.4.

3 See Censorinus, *DN* 22.7 for Greece.

4 Comprehensively in ROSEN 1985; PAILLER 1988: 623–67.

5 Liv. 1.18.2.

The Roman Calendar from Numa to Constantine: Time, History and the Fasti, First Edition. Jörg Rüpke.

quite pure, is nevertheless comprehensible, the conclusion arrived at was that the 'Pythagorean' Roman calendar must have been instituted by Numa 'the Pythagorean'. Marcus Fulvius Nobilior, who lived at the time of the discovery, and, as we shall see, was intensely interested in the calendar, may have been the one who formulated that conclusion, which he then became instrumental in broadcasting.[6] Seen in a historical perspective, of course, the Pythagorean evaluation falls short of the mark. Moralizing and divinatory perceptions of fundamental oppositions such as left-right, straight-crooked, male-female, east-west are common, and in principle arbitrary: which is precisely why the non-arbitrary character of such perceptions is commonly upheld by frequent allusion to metaphorical connections between the body personal and the body social.[7] Whether *impar numerus* actually provides the historically correct explanation in this case must remain unproven, but, in the absence of alternatives, it is at least plausible.

Alluring as this 'Numan' conception was, it was unable entirely to extinguish competing traditions, which have been helped back into the light of modern scholarship by MICHELS.[8] Two second-century BCE annalists, C. Sempronius Tuditanus and L. Cassius Hemina, ascribed an intercalation law to the Second Decemvirate, charged with drafting the Twelve Table Laws.[9] This account ties in with Cicero's firm conviction that *fasti* had been published alongside the law tables. This, however, created problems for him in understanding the publication by Cn. Flavius, one hundred and fifty years later.[10] How are we to deal with these contradictions, which reflect the various Roman reconstructions?

Ovid too attests to the tradition of greater involvement in the calendar on the part of the *decemviri*. Without giving a clear idea of the measures taken, he ascribes to the college the adjustment by which January, from time immemorial the first month, and February, originally the month of the underworld at the year's end, were supposed to have been brought into their familiar sequence.[11] It remains more plausible, however, to connect the reform of the calendar with firm evidence pointing to the publication of the calendar at the end of the fourth century, rather than associate it with sparse, obscure, and possibly fictional accounts of the calendrical activities of the Second Roman Decemvirate, which was a sort of emergency regime in the middle of the fifth.[12] Macrobius,

[6] MICHELS 1967: 122–5.

[7] Taken further in DOUGLAS 1973.

[8] MICHELS 1949: 334ff.; 1967: 119–44 and *passim*.

[9] Macrob. *Sat.* 1.13.21: *Tuditanus refert libro tertio Magistratuum decem viros, qui decem tabulis duas addiderunt, de intercalando populum rogasse. Cassius eosdem scribit auctores.*

[10] Cic. *Att.* 6.1.8 (50 BCE): *ille vero ante decemviros non fuit ... quid ergo profecit, quod protulit fastos? ...* § 18: *Nam illud de Flavio et fastis, si secus est, commune erratum est.* MICHELS 1967: 109–10; sceptical POMA 1984: 275–7.

[11] Ov. *Fast.* 2.47–54. The reference to the passage is from SAMUEL (1972: 166 n. 2), who nevertheless maintains his scepticism towards MICHELS' reconstruction.

[12] I here follow the arguments of HUMM 2000 against MICHELS 1967 and her dating in RÜPKE 1995a; see HUMM 2005: 455–69 for a summary of the arguments. As elsewhere, HANNAH 2005 here follows MICHELS in his datings and interpretations.

our most reliable source, sees Flavius' initiative as marking the end of the lunisolar calendar.[13] The eclipse of the sun on the June Kalends in 344, the day of the dedication of the temple of Juno Moneta, always supposing a dark moon on that occasion, would fit well with the old calendar.[14] Ennius' account of a solar eclipse on some June Nones, on the other hand, while impossible in a calendar that was still lunisolar, certainly provides no proof of the demise of that calendar. In view of Ennius' early dating of the city's foundation, shortly after the fall of Troy, his '350 years later' does not lead to any definable date; what we see here is more likely a reverse calculation using dates from the calendar that applied in Ennius' time.

The peculiar lengths of the months, and the character of the intercalary month indicated in the Fasti Antiates maiores, demonstrate that the originators of the system intended regular intercalation, to take place every two years. The calendar's form implies an intercalary system. The new procedures were of no great consequence. Intercalation was not a new phenomenon. Even the authentic lunar calendar interpolates a month every two or three years, so as to resynchronize with the solar year. The Romans cannot have noticed any great difference.

The measure that had a critical and universal effect on everyday religious, political, and economic life concerned the changes to the months. Their length became fixed, and now distinguished some months from others. Reconstruction of the intercalation procedure[15] demonstrates an association between the month-lengths and the new procedure. The requisite two-yearly intercalation of twenty-two or twenty-three days[16] was effected by inserting the twenty-seven-day intercalary month after 23 or 24 February. The end of the intercalary month corresponded to the five remaining days of February, which thus did not need to be added subsequently, but could be allowed to lapse. A striking characteristic of this structure is that the planners did not retain any authentic lunar months, and must have known at least approximately the correct length of the year, and yet they did not hit upon the comparatively simpler solution found by the Egyptians, who arrived at a year of three hundred and sixty-five days by adding five additional, unattached days to a year of twelve months, each with thirty days, or Caesar's solution, where the existing months were lengthened to make up that same total sum without exceeding the maximum length of thirty-one days. The choice of the much more complicated option suggests that uneven numbers did in fact play a role, and the concern was to preserve the fiction of an entire intercalary month of normal length. The latter point is impressively confirmed by the Fasti Antiates maiores, with their thirteen monthly columns.

[13] Macrob. *Sat.* 1.15.9.

[14] Liv. 7.28.6–8 with Ov. *Fast.* 6.183–5.

[15] Michels 1967: 16–18; see also 145–72.

[16] Censorinus, *DN* 20.6.

4.2 The Character and Significance of the Reform

The view that the Early Republican abolition of the lunar month amounted to a (virtually) unparalleled proceeding[17] has gained in substance. In terms of a conflict between the economic and religious systems, it was an epochal event. The introduction of the *nundinae* would correspond to the removal of our Sunday, or the superimposition of a different cycle of rest-days, as instanced by the *décadi* in the French Revolutionary calendar.[18] We do not know of any rules applying to the occasions where *nundinae* coincided with a festival or an orientation day. Disputes over the legal status of market days[19] suggest that at least the economic function of the *nundinae* was given largely unqualified priority.[20]

The religiously mediated economic (market-related) and political (assembly related) integration and regulation of society in respect of time underwent a process of differentiation. In terms of the calendar, the economy (*nundinae*) and religion (orientation days) formed two separate systems. The gradual formation of an autonomous political and juridical time-structure, which can be seen in Cn. Flavius' *fasti*, and which separated the days of the people's assemblies from both the *nundinae* and the orientation days, may be seen as arising from this disjunction between the days that retained a religious complexion and those where a more economic dimension prevailed. The reassignment of functions to two sequences of days made both categories unfit for the purpose of establishing the highest possible degree of openness: and this at a time, after the tribal reform of 312, when there was a need for precisely such an opportunity.[21]

The concessions falling upon the religious structure, while extensive, were largely hidden. Here, the fiction of the lunar calendar was fully maintained.[22] The *pontifex minor* persisted in his claim to observe the moon. He went to the Arx with the *rex sacrorum*, and proclaimed '*Te kalo Iuno Covella*', five times in January, February, April, June, Sextilis, September, November, and December, seven times in March, May, Quintilis, and October.[23] Seen like this, only one thing changed for the cult: no crescent moon on the Kalends, no full moon on the Ides. Even this had consequences; in some areas these were severe, but for the moment they may be left to one side.

17 For parallel developments see Sacha Stern's forthcoming study on Mediterranean calendars (Oxford University Press).

18 See MEINZER 1992: 111–12, 150–3.

19 See Ch. 5.5.

20 On the other hand, the Jewish calendar for one has a multitude of rules governing conflicts between particular festivals and weekdays (see OBERMANN 1956: 292–3).

21 HUMM 2005: 469.

22 JOCELYN 1966: 97; FLAMANT 1984: 193.

23 'A non-functional rudiment' (GJERSTAD 1961: 197); for an alternative view SABBATUCCI 1978: 461.

Where were further compromises made? In the uneven numbers perhaps, where one explanation may lie in the habit of thinking in *nundinae*. In the inclusive reckoning used, halved *nundinae* produce five-day periods, further halvings three-day periods: the one- and three-day intervals that frequently separate two cult dates in the calendar. As a further element in the adjustment to the new system, the later location of the intercalary point between Terminalia and Regifugium reflects the old intercalary practice. An internal religious 'compromise', in other words a successfully undertaken adjustment, was to establish a nundinal cult, with precedents in the routine cult of the orientation days. The old calendar, in its origins lunar and sacral, was detached from its thematic, astral cult cycle. But the religious structure did not continue to follow its old timetable: it adopted the new calendar. The 'profane' calendar provided the guiding principle for the cult, not the other way round.[24]

So, how did the new calendar look, so far as the part discussed up to now is concerned? The nundinal letters *A–H* – with the letter *Z* in place of the later *G* – were the new element.[25] The letter *G* did not emerge until the third century. *Z* (representing voiced intervocalic /s/) was available instead of *G* until the end of the fourth century. Appius Claudius himself is said to have realized the rhotacization (/r/) of this sound, in written as well as spoken form. Accordingly, the *fasti* of Cn. Flavius may have had *A B C D E F Z H*, the Greek sequence including digamma, as nundinal letters: they were thus responsible for the fact that *Z* did not fall away until *G* emerged to replace it in the set sequence of nundinal letters. Each day was assigned a set letter, beginning with *A* on 1 January. If, in one year, markets fell on days bearing the letter *A*, in the following year – 29 December bears a *C*, 1 January is again an *A* – they fell on days bearing the letter *F*, in accordance with the continuing cycle. Thus, maintaining the eight-day cycle, the relevant letter changed at the end of the year, and when an intercalation intervened.[26] Such a calendar had no need of numbers to indicate intervals: the nundinal letters – like our 'shipping weeks' – allowed rapid orientation, even over quite long periods.

A final, or perhaps initial, consequence of the committal to written form, however, and regarded by the tradition as the central element, was the regulation of intercalation. Differentiation of the nundinal cycle would as such also have been possible within a continuing lunar calendar; but it was, crucially, only the stabilization of the system in written form that created the possibility of planning, in the context of a standard social calendar that was binding on everyone. A further necessary concomitant of the transition to written form, on the other hand, was the need to determine the length of the intercalary month, and its precise coor-

[24] LASKER AND LASKER (1984) demonstrated the divergent development of one and the same ritual in the context of a lunisolar (Israel) and a solar calendar (Babylonian diaspora).

[25] Mart. Cap. 3.261; Pompon. *Dig.* 1.2.2.36.

[26] For a full account Ch. 2.1.

dination.[27] Intercalation itself was no longer subject to administrative decision, as it was now virtually automatic. Within a few decades, a penalty was to be paid for this lack of institutionalization, as we shall shortly see.

Seen in the broader historical context, the abolition of the lunar month and establishment of a purely solar calendar constitutes the most astonishing aspect of the reform. When the Julian year was introduced as a standard in various regions of Greece in the Augustan period, that territory still had empirical or conventional lunar calendars; in this instance, moreover, for months with thirty-one days the first day was counted twice, so as not to allow the overall, notional figure to exceed thirty.[28] Although there are African and Pre-Columbian American examples to the contrary, lunar months dominated religious traditions until the adoption of the Gregorian Calendar in the Modern Age. Islam, with its pure lunar calendar, is something of an isolated case; but lunisolar calendars are characteristic of Chino-Japanese religions, as well as post-Roman Judaism.

We do not have a sufficiently precise picture of developments in the rest of Italy. The so-called Liber linteus of the Agram mummy-wrappings, the most important calendar document for Etruria, can, as such, scarcely be much older than the first century BCE,[29] even if, formally, it may go back appreciably further. In terms of its structure, the document is a *feriale,* a list of dates over twelve months, with detailed although frequently misunderstood descriptions of cult activities ('sacrifice calendar').[30] While it provides information on gods and cult activities, it is of little help for a reconstruction of the technical aspects of the Etruscan calendar, or, to be more precise, of one particular Etruscan calendar. There are no Italic calendars that provide day-by-day documentation, in the manner of the Roman *fasti* (which were also set up in the rest of Italy). Etruscan dates, and such objects as the clay tile from Capua,[31] indicate a system of month-names similar to the one familiar to us from Rome.[32] The incremental counting of all days in the month clearly differs from Roman procedure, and corresponds to Greek practice.[33] As no known monthly date goes beyond the 'thirtieth', there is nothing to indicate that the Etruscans did not retain a lunisolar calendar. The terminology of the dates contained in the Iguvine Tablets also indicates empirical lunar months, in which observation of the first visible crescent moon and the full moon played a part.[34]

27 Cf. GOODY 1968: 35; 1990: 165.
28 LAFFI 1967: 67.
29 EDLUND-BERRY 1992: 331.
30 PFIFFIG 1975: 103–10; RONCALLI 1980, 1980a.
31 See *ibid.*, 422; photo: Pl. 133.
32 EDLUND-BERRY 1992.
33 See SAMUEL 1972.
34 BORGEAUD 1982: 24–33 on *Tab. Iguv.* 2a.15–21 (first crescent); on Umbrian 'Ides' *(plenasier)* 34ff.

5

The Written Calendar

5.1 Gnaeus Flavius

Gnaeus Flavius, a scribe (*scriba*) – probably *scriba pontificius*, the same as the official responsible for the ritual on the Kalends – under Appius Claudius ('Caecus'), was elected curule *aedilis* in 304, against keen competition.[1] While the strain of tradition represented by Livy gives his filiation as *Cn. f.*, Cicero speaks of *Anni filius.*[2] In the latter tradition, it is not the *aedilis* but the *scriba* who published the *fasti* and the *legis actiones.*[3] It may be that this divergence is due to an identification of two different individuals; in that case, the act of publication should be credited to the scribe.[4] It must, however, be said that the chronological component of the dedication on the temple of Concordia, which must necessarily be the work of the *aedilis*,[5] suggests that the two individuals were one and the same. The year of publication thus remains uncertain. It cannot, however, have been appreciably earlier than 304.

The *éminence grise* in this case was called Appius Claudius Caecus. He was censor in 312, and a significant force in political and religious life for the next thirty years, the period when an important phase of both the 'Conflict of the Orders' and the

[1] Liv. 9.46.1–6: *Eodem anno Cn. Flavius Cn. filius scriba, patre libertino humili fortuna ortus, ceterum callidus vir et facundus, aedilis curulis fuit. ... (5) civile ius, repositum in penetralibus pontificum, evolgavit fastosque circa forum in albo proposuit, ut, quando lege agi posset, sciretur; (6) aedem Concordiae in area Volcani summa invidia nobilium dedicavit, coactusque consensu populi Cornelius Barbatus pontifex maximus verba praeire ...* Val. Max. 2.5.2; Macrob. *Sat.* 1.15.9; WOLF 1980a; HUMM 2005: 441–55.

[2] Cic. *Att.* 6.1.8; Plin. *HN* 33.17.

[3] Cic. *Att.* 6.1.8; *Mur.* 25, *De or.* 1.186; Plin. *HN* 33.17; Pompon. *Dig.* 1.2.2.7.

[4] SESTON 1951: 436; BROUGHTON 1951: 168, 1986: 92; Piso, *Hist.* 27 Peter (= Gell. 7.9.2); Macer, *Hist.* 18 Peter (Liv. 9.46.1); Diod. Sic. 20.36.6.

[5] Plin. *HN* 33.19: 204 years after the dedication of the Capitoline Temple.

The Roman Calendar from Numa to Constantine: Time, History and the Fasti, First Edition. Jörg Rüpke.

Samnite Wars was drawing to an end. In his political reforms, he did not flinch from fundamental interventions in religious affairs. He transferred control of the gentilician cult of Hercules on the Ara maxima from the *gens Potitia* to *servi publici*,[6] and excluded the *tibicines* from their regular meeting place, the Capitoline temple of Jupiter Optimus Maximus, where they were in the habit of revelling.[7]

Appius' politics were characterized by centralization and hierarchization,[8] a shift of religious and political authority away from a number of small centres and groups and towards a single, broadly based, substantial and effective centre, located, most importantly, among the urban population. That centre was embodied by the Senate.[9] It is this stance that explains his resistance to the *lex Ogulnia*, as reported by Livy.[10] The opening up of the pontifical and augural colleges, sought for under this law, strengthened both priesthoods, and entrenched them as centres of power independent of the Senate.[11] Any analysis of the publication of the formulae for introducing a civil process before the *praetor* (*legis actiones*), and of the list of days available for doing so (*fasti*), must be set against this background. A desire on Flavius' part to enhance his own profile may be discounted immediately: the driving force was Ap. Claudius, who was already established in the pontifical college.[12]

No further attempt at placing the text politically can be made without first addressing the question of the precise contents of the published *fasti*. Of central importance in this respect were the daily characters. In the extant *fasti*, these are *fas* and *nefas* (with various distinctions of detail) and *dies comitiales*. In practice, the latter also functioned as days on which cases could be initiated before the *praetor*, differing from *dies fasti* only in that, on days marked with a *C*, the magistrates could also summon *comitia*, decision-making people's assemblies.[13] The central criterion by which one day was distinguished from another was the question: *fas*, yes or no? In this alone lies the justification for speaking of the calendar from this point on as '*fasti*'.[14]

If we look closely at the *fasti*, however, the term itself appears curious. *Dies fasti* are in the minority. And this is not all: some *dies fissi*, days such as *Q(uando) R(ex) C(omitiavit) F(as)*, have been given an *F*, when similarly situated days in other

[6] The imparting of cult knowledge at great expense to slaves is supposed to have led to the immediate extinction of the Pinarii and the blinding of Appius: Liv. 9.29.9–11; Festus, *Gloss. Lat.* 270.5–16 L; Serv. *Aen.* 8.179; 269. See WISSOWA 1912: 274–5; a full account in PALMER 1965: 293–308.

[7] Val. Max. 2.5.4, after Liv. 9.30.5–10.

[8] See also PALMER 1965: 320.

[9] STAVELEY 1959: 415–20.

[10] Liv. 10.7.1.

[11] D'IPPOLITO 1985: 125; see also 1986: 103; WIEACKER 1988: 527; similarly STAVELEY 1959: 432; see also HÖLKESKAMP 1987: 62–3; 1988; RILINGER 1976: 97–100.

[12] RÜPKE 2008, no. 1172.

[13] Varro, *Ling.* 6.29–32; Ov. *Fast.* 1.45–54; *Inscr. It.* 13.2.111–3 (Fast. Praen.); Macrob. *Sat.* 1.16.13–14.

[14] Paul. Fest. 78.4–5 L: *Fastorum libri appellantur, in quibus totius anni fit descriptio. fasti enim dies festi sunt.* See also Prisc. *Inst.* 3.486.19–21: *Ipsa res fatus et fatio et fatum. inde putant quidam etiam fas et nefas dictum esse, quod iustum est dici vel taceri. ex quo fasti et nefasti dies et fasti* tà hēmerológia.

months have a *C*. These seeming anomalies, and the legal implications of the two types of day, are explained by the hypothesis that all *C* days were originally characterized by an *F*.[15] For the time being, therefore, we can concentrate on the distinction between *fas* and *nefas*: there will be opportunities later on for discussing these differences of detail.

Taken as it stands, with its entries written in large letters, or in the form to which our discussions have reduced it so far, the calendar's function is essentially the sacral evaluation of non-sacral actions. The *pontifices* may well have been custodians of the law, but the praetorian formulae were not sacral acts. *Nefas* did not arise from an error in performing the cult, such as the wrong form of words, or from the appearance of portents: it denoted a time-limited prohibition, evidently fenced around with sacral sanctions.

What, to refine our inquiry further, was affected by such a classification? The answer must be, by no means legal proceedings as such. The label *nefas* prohibited the presentation of a case before the *praetor* in *legis actiones*, forbade so to speak the activation of the 'interface' between state apparatus and citizen, the process of *cum populo agere*. This, and only this, is implied by the stereotypical reference to the *tria verba*, '*Do, dico, addico*',[16] by which the *praetor* accepted that a case should be presented for resolution by trial: he did not himself try cases. The sources were all written after the differentiation of *dies comitiales* from *fasti*; thus, in discussing *dies fasti/nefasti*, they do not stress that the *comitia* were similarly affected by this fundamental opposition; they too involved communication between the magistrates and the citizens, and so themselves constituted an instance of *cum populo agere*.[17]

What was the purpose of the prohibition of trials and people's assemblies on particular days? There are two possible interpretations. The first may be termed theological: the ruling demands, at an admittedly rudimentary level, a rest from work; the day belongs to the gods.[18] If this had really been the intention, the so specifically formulated rule would have run the risk of being seen as a caricature of itself. However broad (or narrow) one's favoured formulation of the Roman conception of work, much, even in the public realm, would stand beyond its scope.[19] A particular problem would be posed by those entire blocks of days that are marked *nefas*, and which, in particular instances, may suggest a religious motivation: an example would be the first half of June, and the association of the end of that period with the cult of Vesta (*Quando stercus delatum fas*). Other instances, however, definitely do not lend themselves to such an explana-

15 Michels 1967: 39–40, 106–11; Paoli 1954.

16 Varro, *Ling.* 6.29–30; Ov. *Fast.* 1.47–8; Macrob. *Sat.* 1.16.30; *Inscr. It.* 13.2.111 (Fast. Praen. 2. 1.); Festus, *Gloss. Lat.* 162.24–32 L; Paul. Fest. 83.6–7 L.

17 Cf. Liv. 1.19.7.

18 Paoli 1952; Latte 1960: 39.

19 See Michels 1967: 68, 78.

tion, as when, in February, the first half of the month is barred, but the major part of the period of the cult of the dead and the *dies parentales* is marked with a *C*.

The Fasti of Cn. Flavius (Reconstruction)

A K IAN F	F K FEB N	B K MAR NP	A K APR F ...	Z K INT F
B F	Z N	C F	B F	H F
C F	H N	D F	C F	A F
D F	A N	E F	D F	B F
		F F		
		Z F		
E NON F	B NON N	H NON F	E NON N	C NON F
F F	C N	A F	F N	D F
Z F	D N	B F	Z N	E F
H F	E N	C F	H N	F F
A ACON	F N	D F	A N	Z F
B EN	Z N	E F	B N	H F
C KAR NP	H N	F EN	C N	A F
D F	A N	Z EQVIR	D N	B F
E EID NP	B EID NP	H EID NP	E EID NP	C EID NP
F EN	C N	A F	F N	D F
Z KAR NP	D LVP NP	B LIB NP	Z FOR NP	E F
H F	E EN	C F	H N	F F
A F	F QVIR NP	D QVIN BP	A N	Z F
B F	Z F	E F	B N	H F
C F	H F	F F	C CER NP	A F
D F	A F	Z N	D N	B F
E F	B FERA F	H TVB NP	E PARIL NP	C F
F F	C F	A QRCF	F N	D F
Z F	D TERM NP	B F	Z VIN NP	E RECIF N
H F	E RECIF N	C F	H F	F F
A F	F F	D F	A ROB NP	Z EN
B F	Z EN	E F	B F	H EQVI NP
C F	H EQVI NP	F F	C F	A F
D F	A F	Z F	D F	
E F		H F	E F	

A second interpretation, to be described as economic, envisages a contrary intention: the purpose of the lengthy *nefas* periods in the early part of the year and early summer was precisely that farmers should not be kept away from their work by trials or people's assemblies;[20] this interpretation is certainly true for the Imperial-Period *feriae*.[21] However, neither the dates concerned nor the events assumed for those dates are sufficiently convincing. The summer *nefas* periods, from 5 to 15 June and from 1 to 9 July, are rather short for the purpose suggested. Furthermore, the argument of excessive specificity again arises: the absence of trial openings and decision-making people's assemblies would be insufficient to guarantee undisturbed working, when all other stages of trials, together with informative gatherings (*contiones*), were allowed.[22] No quorums were required in order for assemblies to be authoritative; and civil processes, in the final analysis, were the problem of the individuals involved.

The identity of the intended targets of the prohibitions decreed in the calendar is shown by the penalties for infringement. The praetor had to make an 'expiatory sacrifice', a *piacularis hostia*. The parties to the trial incurred no penalty: the opening of their trial was legally valid. The penalty applied, of course, to unintentional infringements (*imprudens*). An intentional violation was inexpiable, would render the magistrate *impius*, and thus – if there was a complainant, whether a malcontent, an enemy, or a rival – deprive him of the legitimate basis for pursuance of his office. How far this distinction extended back over the first century of the pre-Christian era must remain an open question; it did not, in any event, play any decisive role. There is no mention in the tradition of a magistrate's being dismissed for intentional infringement of this rule: in a doubtful case, we may assume that the expedient of pleading negligence was always open. The mechanism by which the sanction operated was more subtle: even in the case of an unintentional error, the *de iure* valid action remained tainted by *vitium* under sacral law, and thus rendered the magistrate in question vulnerable to political challenge. Under the conditions of acute rivalry that characterized Rome's political system from the Middle Republic onwards,[23] a magistrate elected for only one year could not countenance such exposure; although, in fact, we know of no instances of such a charge being made.[24]

Critically affected by the rule was the magisterial *ius cum populo agendi*, a *ius* that grew out of the office itself, and was here restricted by a predication of *fas* or *nefas*. The law involved related to transferred authority (*potestas*), and deter-

[20] MICHELS 1967: 116; *Dig.* 2.12.1, pr.

[21] See e.g. *Cod. Iust.* 3.12.5.

[22] Cf. Tac. *Ann.* 4.70.3.

[23] See RÜPKE 1995b.

[24] Varro, *Ling.* 6.30: *Quod si tum imprudens id verbum emisit ac quem manumisit, ille nihilo minus est liber, sed vitio, ut magistratus vitio creatus nihilo setius magistratus. praetor qui tum fatus est, si imprudens fecit, piaculari hostia facta piatur; si prudens dixit, Quintus Mucius aiebat eum expiari ut impium non posse.* Cf. 6.53; *Lex Irnit.* 10 B 36–50; Macrob. *Sat.* 1.16.10; Festus, *Gloss. Lat.* 234.3–5 L.

mined mutual power-relationships between magistrates.[25] In respect of the balance of power within the aristocracy, it was not trials but the *comitia* that were of central significance, formally speaking. This is sufficiently demonstrated by the greater number of the later *dies comitiales* as against *dies fasti*. Time-related restrictions in respect of this 'right to treat with the people' had political implications. Increasingly, the assemblies took on the nature of bottlenecks within the process of political communication and decision-making; the *dies comitiales*, by their availability and their immunity from the hindrances occasioned by moveable feasts, acts of expiation, games, and replays of games, formed the 'window' through which a judgement could be implemented.[26]

There were two components to the function of assembly-free days in the system of political institutions. Sittings of the Senate could take place on such days. The list of Late Republican Senate sittings for which we can determine dates[27] shows that *dies comitiales* were only rarely used for this purpose. Some 72 per cent of sittings are attested to have occurred on other days, which made up only 42 per cent of the year.[28] Most of the exceptions represent days immediately preceded by 'regular' sitting-days. The Senate could meet on any category of day, but encroached on assembly days only when compelled by circumstance.[29]

The second function of *nefas* days remains hypothetical, and turns the 'economic interpretation' on its head: the periods free of assemblies offered the opportunity for magistrates to take a 'holiday'. They restricted the possibility of colleagues' taking unfriendly initiatives, capable of being thwarted only by personal attendance and use of a veto. The dates of Senate sittings fail to provide significant confirmation for this hypothesis. Periods for which we have no record of sittings coincide with the picture given by the calendar in April[30] and at the beginning of July. February, on the other hand, an explosive month in respect of foreign affairs, falling prior to the beginning of spring, when thoughts turned to military activity, bristles with Senate sittings, in spite of lavish marking of the calendar.

Examination of the distribution of court and assembly days reveals a further circumstance: *comitia* were prohibited on the Kalends and Nones themselves, and on the days following. The 'natural' assembly days, those dates on which the city should have been frequented for other reasons, remained barred for the purpose of political decision-making. After the separation of Nones from ***nundinae***, this may appear bewildering: what other practical function did the Kalends and Nones possess? Our bewilderment is lessened by the realization that the same *lex Hortensia*

25 See HEUSS 1944; BLEICKEN 1981; HEUSS 1982; RÜPKE 1990a: 41–51.

26 MICHELS 1967: 36–50; see also RILINGER 1976: 94–105; BURCKHARDT 1988: 178–209; LIBERO 1992: 53–68; RÜPKE 2007.

27 MICHELS 1967: 55–60; BONNEFOND-COUDRY 1989: 199–229.

28 BONNEFOND-COUDRY 1989: 220.

29 Special rules during the final decades of the Republic are fully discussed by the above two works, and need no more detailed coverage here.

30 See TALBERT 1984: 209–10 for Senate 'holidays', particularly in April/May.

to which may be attributed the differentiation of the *dies fasti* from the *comitiales* also, in 287, gave the *nundinae* the character of *dies fasti*: in short, days without decisions of the people.[31] The character previously possessed by these days is unknown.[32] There is much, in the discussion concerning the *feriae* status of the *nundinae*, and the *religiosus* rulings in respect of the orientation days and the days following, to suggest that Flavius classified these days as *nefas*.[33] Only the differentiation between *fas* and the *comitialis* function would permit them to be qualified as *fasti*.

To return to Flavius: what did his accomplishment actually amount to? He published a list of *fas* and *nefas* days, imposing restrictions of time on the actions of magistrates. This does not bring us appreciably beyond the information given by the sources, even if we have gained a more precise understanding of the regulatory profile. The aspect passed over in silence by the sources on Flavius is fine differentiations within the *nefas* category, such as *QStDF* and *EN*, designations that are at least partly clarified by systematic descriptions of the calendar by antiquarians such as Varro and Macrobius. These authors bring us closer to the character of Flavius' list. The key to an understanding of its essence, however, is provided by an abbreviation that is as mysterious as it is self-evident, although it does not appear to have been clarified at all by the ancient sources, and has not been conclusively resolved by modern scholarship: *NP*.

5.2 *NP* Days and Feast-names

The religious character of the *nefas* designation becomes clear in the sanction of *piaculum*. This term applied equally to the offence (*piaculum est*) and the 'expiatory sacrifice' it gave rise to (*piaculum facere*). Given the etymology from *pius* and *-culum*, the term describes the instrument that is capable of restoring the condition of *pius* after the offence.[34] In antiquarian passages of a technical nature, equal status nevertheless appears to be claimed for the first meaning, the offence itself. *Piacularia auspicia* were ill-omened auspices that emerged at the sacrifice, and could thus presumably be defused by *piaculum*;[35] *sine scelere, sine piaculo* denoted the permissibility of an action, just as *piaculum esto* expressed the contrary judgement.[36] The semantic association depends on the inevitability of the acts that restore *pietas*: the offence against sacral law (*piaculum esto*) allows of no choice; it must be cor-

[31] Macrob. *Sat.* 1.16.30.
[32] MICHELS 1967: 106: *fas*.
[33] See the cited passage from Macrobius and RÜPKE 1995a: Ch. 15.3.
[34] TROMP 1921: 26–30.
[35] Paul. Fest. 287.3–6 L; cf. TROMP 1921: 34–5.
[36] *CIL* 1².756, Z. 10 = *ILLRP* 508 (temple law from Furfo); Varro in Macrob. *Sat.* 1.16.20.

rected (*piaculum dato*).[37] Here again, it must be emphasized that offences explicitly in the context of sacral *law* were made subject to sanction by being qualified as *piaculum*.

The debate referred to by Macrobius, concerning the religious evaluation of particular days,[38] clearly distinguished between strict prohibitions, applying to *feriae*, and those days on which something *religiosum est*, or *non oportet*, that is to say should be avoided, or was inappropriate. Two passages cited from our main source in this context, Marcus Terentius Varro, illustrate this differing evaluation. The first passage comes after a lengthy discussion of *dies proeliares*, on which it was permissible to fight against the enemy; it mentions the concepts *fas* and *nefas*, but not *piaculum*, and concludes with the keyword *religiosum*:

'When the *mundus* is open, the door of the gods of misfortune and the underworld may also be regarded as being open; this is why it is irreligious, on those days, not only to engage in combat, but also to recruit soldiers or send them to the army, to set sail, or to join with a woman in lawful matrimony for the sake of procreation.'[39]

In the passage directly following, cited by Macrobius from a work by Varro concerning the augurs, *feriae* is mentioned for the first time, and here we find the keyword *piaculum*, in correction of a *non oportet* felt to be too lenient:

Viros vocare feriis non oportet: si vocavit, piaculum esto ('Men should not be summoned on holidays; if (the magistrate) calls, a *piaculum* should be held').[40]

Viros vocare functions as a synonym for the summoning of the *comitia centuriata*, the people's assembly, drawn up in their fighting divisions: the *exercitus urbanus*.[41] We already know of a similar prohibition of assemblies in respect of *nefas* days. Here, though, the talk is only of *feriae*: they call for a *piaculum*. But not all *N* days were *feriae*.

It seems appropriate to turn back to the calendar here. Our perceptions now sharpened, we are struck by the difference between the notes *N* and *NP*. *NP* denotes *feriae*: this is not only a matter of *communis opinio*, but is made explicit in many calendars by means of captions. The conclusion to be drawn from our observation is that *NP* is short for *N(efas) P(iaculum)*.[42]

The objection is immediate and justified: in their explanations of *dies nefasti*, our sources associate them with the *piaculum* ruling without exception. If our

[37] See e.g. the sanctuary laws from Spoleto and Trevi (*ILLRP* 505–6): *seiquis scies violasit dolo malo, (et*, Trevi) *Iovei bovid piaclum datod (dato*, Trevi).

[38] Macrob. *Sat.* 1.16.

[39] Macrob. *Sat.* 1.16.18.

[40] Macrob. *Sat.* 1.16.19.

[41] See RÜPKE 1990a: 35ff.

[42] MICHELS (1967: 68–83) also points to the close association between *feriae* and *NP* days; her interpretation, however, follows SOLTAU 1889, and after him WISSOWA (1912: 438) (*nefas: feriae publicae*), to arrive at *(dies) nefasti publici* (1967: 76; on the research history 74–6). The attempt by PAOLI (1950) to differentiate *N(efas)* and *N(efas: feriae) P(osteriores)*, in respect of those festivals marked only with *N* (258–60), from a more recent group of 'later festivals', to be assigned to the pre-Capitoline Triad, must be seen as having failed: here too, PAOLI does not

interpretation of *NP* were correct, surely all *N* days should be marked, not *N*, but *NP*. Or, conversely: if no *piaculum* could be incurred on *N* days, why are they marked at all with this abbreviation?

The religious element does not lie in the *N*, but solely in the *P*; this alone denotes the point at which the *pontifices* remain involved, in their function as *piaculum* specialists.[43]

The effect of the list published by Flavius lay in its application of the calendar to the regulation of political activity. As in the case of the Twelve Tables, we may assume that the codification did not provide any revolutionary new design, but confirmed an existing situation, perhaps with relatively slight corrective and systematizing touches. In this instance, the major concern was the practical question as to the days on which *comitia* were held, and those on which they were not. The grounds and motives for this practical concern may have been many and various; reasons of strictly sacral law, defining genuinely sacrosanct days, or *feriae* (*NP*), existed alongside vaguer religious considerations, whereby particular cult events were accompanied by the barring of assemblies, without the option of imposing an actual religious sanction (*piaculum*); but these occasions may also have been accompanied by profane considerations, such as the 'magistrates' holidays' (*N*) already postulated. How was such a practice represented as binding? How were description and custom transformed into a norm?

Evidently, the *fasti* took up a line of reasoning that solved the problems by blurring distinctions. Flavius – and, where Flavius stands, Claudius should be assumed – marked all days *nefas* or *fas*: this measure, which went far beyond the sacral, was legitimized by a religious code, a sacral terminology. This was the first step. And the second: the more detailed the portrayal and justification of the pontifical rules, the more credible was the overall construct, the higher the degree of acceptance. And the entire affair was conducted within the pontifical college.

The strictest degree of differentiation corresponds to the strictest sacral sanction: *feriae* bear the warning *P(iaculum!)*. The characterization is not merely recorded, but also legitimized, in the form of an abbreviation referring to the underlying cult provision. Why an abbreviation? Analysis of Imperial-Period *fasti* has produced a few figures as regards their size as inscriptions. In Flavius' calendar, orientation days and nundinal letters for 13 months, totalling 355 plus 27 days, already required rather more than 500 characters. The addition of daily characters alone virtually doubled this number. If the desire was that the calendar should

address the problem of those *N* days that have no festival abbreviation; the interpretation used relies on K. O. MÜLLER's reading of Festus' fragmentary lemma '*Nefasti dies*'. The extant text breaks off after *addico nep*, and the next line begins with *riores sunt* (Festus, *Gloss. Lat.* 162.27–8 L). MÜLLER reads from this *[... do dico] addico. NP [... qui nefasti poste] riores sunt* (165 M). In view of the lack of parallel instances, and the absence of a convincing concrete explanation, this interpretation is inadmissible. The most recent attempts are scarcely more helpful: *nefastus purus dies* (BRIND'AMOUR 1983: 227). RADKE (1990: 18) returns to the old solution *nefas principio/parte diei*.

43 On this functional area see WISSOWA 1912: 515 n. 1.

be easy to copy, restraint was indicated. In practice, this meant abbreviation. Abbreviations comprising the simple omission of a word's final letter or letters are attested in dedicatory inscriptions of the end of the fourth and beginning of the third century BCE: thus contemporaneously with our calendar.[44]

But the extreme abbreviations of the *fasti* indicate that we should also consider the other aspect of the process: oral familiarity with the feast dates. Seen from this point of view, the *fasti* must have been no more than an aid to memory. The high degree of abbreviation, so unusual in other texts of the same period, and in those Greek inscriptions that we know of, was appropriate for a calendar whose ciphers, in appearance resembling abbreviations, in fact, for the most part, were none (nundinal letters), and the rest of whose words were either short (*eidus, nonae*), or, as dates, familiar in abbreviated form (*K. Febr.?*).

The use of abbreviations does not have a merely palaeographic dimension. In terms of the history of religion, the question arises: what was abbreviated? In retrospect we may say, of course, the feast names. *Vinal* must stand for Vinalia, *Tubil* for Tubilustrium, *Regif* for Regifugium. But does not *Fur* stand as well for the goddess Furrina as for Furrinalia, her holiday; likewise *Car* as well for Carmentis as for Carmentalia? And would this make a difference in respect of the utility of the *fasti*? The significant point lies one stage further back: Flavius needed unambiguous abbreviations; formulations such as '*feriae Furrinae*' were entirely unusable so far as Flavius was concerned, although not for the *rex sacrorum* when he was making his announcements.

All this would be immaterial, and has indeed been so in the eyes of scholarship up to this point; but the new historical reconstruction has given rise to the suspicion that Flavius' inventive enterprise may not have been confined to the daily characters. The fact is, that, prior to Flavius' publication, there was no such call for succinct, unambiguous names for festivals, beginning as far as possible with different letters. It is, further, the case that, in at least one instance to our certain knowledge, the terminology used by the *rex sacrorum* in naming dates did not coincide with the names in the *fasti*. What the latter label *LUPER*, and antiquarian texts refer to as *Lupercalia*, the *rex sacrorum* called *dies februatus*, a name that did not easily lend itself to unambiguous abbreviation.[45] And, in fact, other differences are apparent in abbreviations for festivals and festival names: Quinquatrus or Quinquatria, Agonalia or Agonium, Liberalia or Agonium, Divalia or Angeronalia.[46]

The calendar-related festival names that have come down to us in texts commenting on this very calendar fall into several groups. *Agonium*, for example,

[44] See *ILLRP* 16–17. See also JOHNEN 1911: 150–75, especially 156 on the first-name abbreviations that arose before the third century BCE.

[45] Varro, *Ling.* 6.13: *Lupercalia dicta, quod in Lupercali Luperci sacra faciunt. rex cum ferias menstruas nonis Februariis edicit, hunc diem februatum appellat. februm Sabini purgamentum, et id in sacris nostris verbum: nam et Lupercalia februatio, ut in antiquitatum libris demonstravi* (cf. Serv. *auct. Aen.* 8.343ff.). After Varro, Paul. Fest. 76.1–2 L. Cf. oral *Larentalia* Varro, *Ling.* 6.23; Plut. *Rom.* 4.5.

[46] See Suet. *Dom.* 4.4; Paul. Fest. 9.15–23 L; Varro, *Ling.* 6.14. Angeronalia instead of *DIV*: Varro, *Ling.* 6.23.

certainly featured in the official priestly lists[47] (on whose aspect and extent at this period[48] we must not speculate). It occurs frequently in the *fasti*, but without differentiation in respect of any divine patron. The days in question do not appear to have enjoyed any great degree of public awareness. The formation with *-ium*, plural *-ia*, extended form *-alia*, is already apparent here; word formation is in this respect standard for festival names. An exception is the Quinquatrus ('the fifth day after the Ides'?), whose etymology appears to be no less complex than its cult content.[49] Of the latter, the name reveals nothing, which gives it a certain affinity with the Agonium.

On the basis of word-endings, Regifugium and Poplifugia also belong together, as do Tubilustrium and Armilustrium, and Fordicidia too. In their nominal style, all the formulations evoke striking cult elements: 'flight' of king or people, 'purification' of trumpets or weapons, 'killing' of pregnant animals. The descriptions rely entirely on externals. Why the king (originally the real king, subsequently the *rex sacrorum*) fled, why the people did likewise: everyone has produced different answers. The connection between the Fordicidia on the one hand, and the Cerialia and Parilia on the other, is not apparent from their names alone. 'Purification of the trumpets' represents a false interpretation; the same goes for Armilustrium. To summarize: all are lay terms rather than specialist pontifical terminology.

How is it with the other names, for the most part ending in *-alia*? They conceal a colourful hotchpotch. Faced with this potpourri, let us call on Varro, our best authority on the subject, in the assurance that he more often scores a hit than a miss. *Lupercalia dicta, quod in Lupercali* (location) *Luperci* (priests) *sacra faciunt … Ecurria ab equorum cursu* (horserace: ritual) … *Cerialia a Cerere* (divinity), *Vinalia a vino* (cult purpose and cult material?).[50] What cannot be denied the terms is their transparency: the root element can be recognized and interpreted; which is precisely why these terms above all appear in antiquarian texts.[51] There are only a few exceptions. The Lupercalia offer many possibilities: a location? priests? even a god Lupercus or a Juno Luperca? Uncertainty is the only safe reaction to the frankly incoherent derivation of *Meditrinalia* from a toast: did a goddess Meditrina preside over this festival of young wine?[52] Even these exceptions, however, present no real difficulties: each possibility remained unambiguous. On the other hand, what are we to make of the priestly jargon word *agonia*? The terminology of the *fasti* did not make great demands on the user's ingenuity: that the Furrinalia rep-

[47] Varro, *Ling.* 6.14: Salia, Sab. *Macrob. Sat.* 1.4.15 (*fast.* 2, fr. 9 Huschke): *pontifices*.

[48] Unlikely to be much earlier than Flavius: RÜPKE 1993b.

[49] Name: WACKERNAGEL 1924; WALDE AND HOFMANN 2.408; ERNOUT AND MEILLET 558; see also HENSCHEL 1963: 1149–53; SABBATUCCI 1988: 106. Ritual: USENER 1875: 225–8; ALTHEIM 1932: 1778–9; HERMANSEN 1940: 68–70; HENSCHEL 1963; TORELLI 1984 (with the review by AMPOLO 1988); CAMASSA 1984.

[50] Varro, *Ling.* 6.13–16.

[51] Varro, *Ling.* 6.1–2.

[52] Varro, *Ling.* 6.21; Paul. Fest. 110.21–5 L. Cf. Minervalia (Tert. *De Idol.* 10).

resent *feriae Furrinae* strikes one immediately. In terms of sacral law, it appears equally valid that the Ides were *feriae Iovis*. The critical difference is that the latter information cannot be obtained from the abbreviation *EID*, or from *eidus*. This proves that we are entitled to grasp the other end of the stick: the secondary form is not *feriae Furrinae*, but *Furrinalia*. I prefer not to see this as specialist terminology, but rather as a popular corruption based on pontifical jargon. Insofar as this may not yet have produced the degree of completeness demanded by the new calendar, a further possibility remains: a portion of the Roman feast names that are so familiar to us were invented in the context of the new discourse occasioned by the writing of a calendar.

5.3 Cultic and Linguistic Details

The picture has become more obscure than Flavius deserves. The informative content of 'his' *fasti* – and thus their credibility – goes beyond the marking of *feriae*. Abbreviations are not confined to *NP* days. Some *N* days too, such as the Lemuria, are given 'validation'. The Regifugium also falls into this category, as, perhaps, do the Matralia.[53] In all instances, the dates in question were well-known, even 'popular'. With the exception of the Regifugium, they already had the character of *dies religiosi*. Flavius brought the religious validation to the fore, marking these days *N* 'for safety's sake', even when they were not *feriae*. Flavius' document had the effect of giving the norm *de iure* validation: under sacral law, an act might be an offence against a *religio*, without constituting a *piaculum*. *De facto*, pontifical rules below the level of *piaculum*, combined with the political practices of the nobility, may already have prevented the day's use.

At first sight, the note *FP* (*Fas Piaculum*?) appears contradictory, but the reference is soon resolved. The *P* does no more than indicate a problem for which the *feriae* designation does not suffice. As the note is associated with the Vinalia, it may be surmised that what is involved here is a topographically constrained danger: the name Vinalia rustica on 19 August demonstrates that there was a role for cult acts at the site of cultivation.[54]

We cannot tell from the calendar what part religion played on the other *N* days that bear no further label. Here, I can only refer to the above considerations. It may be possible to establish an association between the *nefas* period in the first half of February, stretching across two *nundinae* from 1 to 17 February, with later rules relating to the Senate's foreign-policy role: deliberations on questions of war

[53] See MICHELS 1967: 184.

[54] See LATTE 1960: 75–6.

and diplomatic tactics were not to be torpedoed by a bellicose magistrate manipulating the people's assembly.[55]

The further, rare categories of *dies fissi* and *dies intercisi* betray a deliberate delight in detail. First, the abbreviation *EN* is usually resolved in modern research as *endotercisus*, which is linguistically[56] and functionally impossible: the daily characters had above all to fulfil the one function of showing at a glance whether a day was designated *fas* or *nefas*. In the literary sources, these days are always cited as *dies intercisi*.[57] The appropriate clarification at first occurrence of the letter sequence in the Fasti Praenestini (on 10 January) fell victim to the dilapidations of time: at the critical location, only *[...] pro in ponebatur* is now legible.[58] The integration that springs to mind in linguistic-historical terms is *[endo] pro in*, but neither as **endoter-* nor as **enter-* is this reconcilable with *intercisus*. The way out of the impasse is to distinguish logically between the genres of the sources: the telegraphic style of the *fasti* need not necessarily be reflected in antiquarian literary accounts. By using two archaic words ultimately reported by Verrius Flaccus, the circumstance described by Varro as 'in the morning and the evening is *nefas*' may be represented in a form capable of being abbreviated to *EN*: *E(ndoitio) E(xitio) N(efas)*.[59] The reduction of the duplicated *E*, so that the force of the second is retained so to speak in ligature in the first, is paralleled in classical epigraphy only in the reduction of *NON(ae) N(efas)* to *NON*; but, in terms of abbreviations, Flavius' project at the end of the fourth century was breaking new ground.

What does the designation *EN* tell us? Did it enable senators to stay longer in bed in the morning, or to retire earlier at night? Probably not: the note had as little practical significance as the days characterized in the literature as *dies fissi*, a collective term for the complex abbreviations *Q(uando) R(ex) C(omitiavit) F(as)* and *Q(uando) St(ercus) D(elatum) F(as)*, which allowed the magistrates *cum populo agere* only after the conclusion of certain rituals. The reason was the time taken for information to be relayed: at a period when there were no accurate clocks, how could time be synchronized at different locations, when overlaps might be measured in minutes? It certainly took less time to kill the sacrificial animal in the morning than to make the court ready for business.

[55] See MICHELS 1967: 43.

[56] See HERMANS 1964. In detail: RÜPKE 1991.

[57] Varro, *Ling.* 6.31: *Intercisi dies sunt per quos mane et vesperi est nefas, medio tempore inter hostiam caesam et exta porrecta fas.* Here as elsewhere, I reject the traditional *lectio facilior (proiecta)* – also Cic. *Att.* 5.18.1 – in favour of the *terminus technicus (porricere)* offered by the traditions of Plautus (e.g. *Pseud.* 266) and Macrobius. Macrob. *Sat.* 1.16.3: *Intercisi in se ... dividuntur: illorum enim dierum quibusdam horis fas est, quibusdam fas non est ius dicere. nam cum hostia caeditur fari nefas est, inter caesa et porrecta fari licet, rursus, cum adoletur non licet.* Cf. Ov. *Fast.* 1.49–52: *Nec toto perstare die sua iura putaris: / (50) qui iam fastus erit, mane nefastus erat; / nam simul exta deo data sunt, licet omnia fari, / verbaque honoratus libera praetor habet.*

[58] *Inscr. It.* 13.2.113.

[59] See Paul. Fest. 67.2 L: *Endoitium initium.* Cf. 71.7–8 L: *Exitium antiqui ponebant pro exitu; nunc exitium pessimum exitum dicimus.* From this can be derived the following integration for *Inscr. It.* 13.2.113 (after RÜPKE 1991): *EN. Haec nota signif[icat diem endoitio exitio nefastum. Nam endo] pro in ponebatur, [exitium pro exitu. His diebus nefas est mane ante]quam hostia immol [etur ...].*

As well as the niceties of the cult, also deserving of attention are the elemental concepts *fas* and *nefas*, and the synthetic terms *dies fasti* and *dies nefasti*, which were derived from the calendar, but not necessarily by Flavius himself. *Fas* and its derivate *nefas* appear in Latin literature from Plautus onwards.[60] There is no doubt that they had a religious character, in the sense of divine and/or moral sanction for actions qualified as *fas*;[61] but did they really comprise generally applicable 'taboos in respect of persons, objects, time, and actions' under sacral law?[62] The sheer number of the references cannot conceal the fact that neither term is ever used in passages of sacral law, or in priestly terminology in the narrower sense.[63] Like *fastus* and *nefastus*, they have a specific application only in relation to the calendar. Notwithstanding the very real implied political and juridical consequences (which, however, reside in the negative *ne fas*), no synonym in sacral law can be found for this calendar-related usage, nor even a clear concept of it formulated.

What is the origin of the term *fas*? Etymology, while providing no further help, does confirm the impression already gained. While it is possible to associate the unambiguously sacral terms *fēriae*, *fēstus*, and *fānum* with a root **fēs* or **fās*, which is still discernible in other Italic languages,[64] *fās* cannot be related to this group, even though such a relationship was established in Antiquity by means of a common derivation from *fari*, 'to speak'.[65] Although *fari* represents a popular etymology,[66] that alone is no reason for regarding it as false. ÉMILE BENVENISTE has shown that *fāri* and its Latin derivatives and Indo-Germanic parallels (*phēmi*, **bhā-*) denote less the act of speaking than the existence of the utterance and its effect.[67] *Fas* may thus be understood as 'that which is permitted by divine edict'. The term in itself implies no specification of the content of the permission implied, nor grounds for (or, in the case of *ne fas est*, against) it.[68] Precisely this openness allowed the extensive use of the terms in the calendar, from which they subsequently acquired a specific connotation.

That application of the opposition *fas-nefas* to all days of the year represented over-systematization is demonstrated by early reaction in the *lex Hortensia*. As will be shown in the next chapter, this law corrected the detail of some rulings in respect of the *nundinae*, and, by introducing the classification *dies comitiales*,

60 Plaut. *Asin.* 514; Acc. *Trag.* 157, 585.

61 See LATTE 1950: 336.

62 WIEACKER 1988: 315.

63 Liv. 1.32.6 reports a fictitious ritual.

64 WALDE AND HOFMANN 1.453–5. 481; ERNOUT AND MEILLET 1.215–16, 226–7. *Fanum* from *fasnom*.

65 The tentative suggestion in ERNOUT AND MEILLET (1.217) that *fās* might belong to **fas* is purely hypothetical: see the critique in WALDE AND HOFMANN 1[4].458; BENVENISTE 1969: 134; but cf. Paul. Fest. 78.6–7 L: *Fanum a Fauno dictum, sive a fando, quod dum pontifex dedicat, certa verba fatur.* 83.5 L: *Fana, quod fando consecrantur; Not. Tiron.* 45.11–17.

66 See also Paul. Fest. 78.26 L.

67 BENVENISTE 1969: 133–40.

68 BENVENISTE 1969: 139–40.

removed the possibility of recourse to *comitia* from the remaining *dies fasti*, thus undertaking a radical redefinition of the concept *dies fastus*. This would scarcely have been conceivable if the institution had been long-established.

5.4 The Purpose of the *fasti*

In the light of the tendency, in ancient and modern narratives alike, to give undue emphasis to Flavius' personal role, his work must not be misunderstood as comprising a secretly prepared copy of pontifical records. It must primarily be seen as a codification of existing practice. The committal to written form was directed against the arbitrary exercise of pontifical power. It restricted the political latitude enjoyed by the *pontifices* through their opportunities to intervene in the political timetable. At the same time, in casting the authority of the traditional rules into written form, it safeguarded that authority. For the *pontifices*, the proceeding may have had a welcome as well as an unwelcome aspect.

Behind these intentions lay the constitutional and political programme of Ap. Claudius Caecus, who, it must be remembered, was himself a *pontifex*. What was envisaged was not a complete separation between the political and religious systems, and by no means the disempowerment of religion. More modestly, codification was to be the means of stabilizing and institutionalizing influences and interdependencies between the two systems. Publication imposed an increasing need for the *pontifices* to justify their more far-reaching conventions, and, to a large extent, that need became an element of political discourse. The political and structural conditions delineated here represent realistic assumptions in respect of the final years of the fourth century, when a new patrician and plebeian nobility was being formed; they are by no means backward projections of the circumstances of the Late Republic. The contemporary *lex Ogulnia* confirms that the priesthoods were in fact perceived as politically relevant entities.

The divergent approaches to the situation, as exemplified in the *lex Ogulnia* on the one hand and Flavius' measure on the other, correspond to two traditions of political action that permeated the Middle and Late Republic. Access to the priesthoods constitutes one of these strands. The social origin of candidates (patrician or plebeian), the appointment procedure (cooption or election), and the size of the colleges, provided the material for numerous legislative initiatives, from the *leges Liciniae Sextiae* to the definitive reorganization by the future Augustus in the years after Actium. The second strand may be traced back to the *fasti*. It comprised repeated attempts, by means of codifications and legal stipulations affecting the interfaces between the religious and political systems, to restrict the freedom of the autonomous priestly colleges to intervene in the political process.

The tendency of the latter tradition, with its inherent (desired) effect of compelling broad consensus within the leading elite, was to militate against the segmentation of power. The outcome is demonstrated by the history of *obnuntiatio*, the augural power of veto, in the Late Republic, and of the use of omens with the effect of veto: in the constitutional context described, the strategy of instrumentalizing the priesthoods trumped any move to disempower them.

As I have frequently emphasized, the *fasti* reflected existing practice. Changes were not entirely out of the question, but they had to command consensus. With one exception that will be addressed shortly, there is no supporting evidence for such revisions in the Flavian *fasti*, that is to say in the tradition that used large letters, for the period of the Republic up to the Julian reform, although the possibility that some such took place cannot be excluded, at least for the third century. Fundamental changes in the character designations of days did not enter the repertoire of devices used in the religious and political sphere until the final years of the Republic; this does not apply to ongoing alterations to the nature of days occasioned by the placement of rituals, the reporting of prodigies etc.

Two factors may have been at work here. In the critical initial phase, the codification as such appears to have been judged controversial, but not its subject-matter, which was unassailable from the standpoint of religious law; politically, the majority of the leading elite benefitted by it. Another element was the document's religious character. Although it greatly simplified and systematized complex matters of religious law and politics for political purposes, the terminology it used belonged entirely to the religious sphere. Neither Cn. Flavius nor Ap. Claudius Caecus may have challenged the rumour that the entire list had been copied from confidential pontifical documents. It was precisely this aspect of its unexplained derivation from supposedly arcane sources, cast, moreover, in what was at least in part a new terminology, that gave the document its aura of authenticity. It acquired a dignity that protected it from the interference of all sections of the body politic, until it met its nemesis in the shape of new sources of authority uncapped by the monarchy of Caesar and Augustus. Until that time, the *fasti* were accepted as the framework for political activity. Only the actual rules governing the different types of day presented openings for any kind of innovation. Such opportunities gave rise to the *leges* forbidding the Senate from meeting at certain times, or declaring that particular types of assembly were permitted on other days.

5.5 The Law of Hortensius

As an instance of what an intervention in the conventions fixed by the *fasti* might have looked like, we have only the *lex Hortensia* of 287. The question posed was, what is the status of the *nundinae*? We are able to reconstruct the original problem, at least to some extent, from the comments of later antiquarians, as cited in some detail by Macrobius.[69] The controversy focussed on alternative views as to whether or not the *nundinae* were *feriae*. Some of the chains of evidence prove to be on the lengthy side. According to Julius Modestus, when the augur Messalla posed the question to the *pontifices*, he received the answer that they did not regard the *nundinae* as *feriae*. On the other hand, it was known that Jupiter regularly received a sacrifice from the *flaminica* in the Regia on those days: a good argument, taking the Ides as a parallel (although an unstated one), for sharing Granius Licinianus' assumption that the days in question were *feriae Iovis*.[70] During the Late Republic, lawsuits could be initiated on these days, but *comitia* could not be held. This corresponds to the Late Republican rules for *dies fasti*, and, according to the information and resolution related by Macrobius, the *nundinae* were designated as such by the above-named law. But this did not solve the problem from the standpoint of sacral law.

For the sake of clarity, let us cast our minds back. Our analysis of the calendar reform revealed the *nundinae* to be artificial creations that took over the function of the orientation days, and, on their model, were given sacral features. But the sacral character of the orientation days themselves was not clear-cut. The Ides had the character of *feriae*; the 'Tubilustria' attracted other *feriae*, but did not of themselves provide a basis for that status: their function as *comitiales* belonged to the day following.[71] The Kalends and the Nones were different cases: assemblies of an informative character were held on both days; the Nones appears even to have lacked any further cult act. Here too, however, we need to be more precise. The *fasti* provide only a partial picture of a complex religious reality. The regular sacrifice to Juno on the Kalends corresponds with Macrobius' assertion that these days too were *feriati*.[72] That no *feriae publicae* in the strict sense of the term can have formed the basis for Macrobius' classification of the Kalends is proved by the *feriae Marti* on 1 March: these would have been impossible if the Kalends were already *feriae* of Juno; it follows, then, that the term *feriatus* could be used in a merely analogous sense. Cato tells us how matters stood from the perspective

[69] Macrob. *Sat.* 1.16.28–30: *Quod autem nundinas ferias dixi potest argui, quia Titus de feriis scribens nundinarum dies non inter ferias retulit sed tantum sollennes vocavit, et quod Iulius Modestus adfirmat Messala augure consulente pontifices, an nundinarum Romanorum nonarumque dies feriis tenerentur, respondisse eos nundinas sibi ferias non videri, et quod Trebatius in libro primo Religionum ait nundinis magistratum posse manu mittere iudiciaque addicere. (29) sed contra Iulius Caesar sexto decimo Auspiciorum libro negat nundinis contionem advocari posse, id est cum populo agi, ideoque nundinis Romanorum haberi comitia non posse. Cornelius etiam Labeo primo Fastorum libro nundinas ferias esse pronuntiat. (30) causam vero huius varietatis apud Granium Licinianum libro secundo diligens lector inveniet. ait enim nundinas Iovis ferias esse, siquidem flaminica omnibus nundinis in regia Iovi arietem soleat immolare, sed lege Hortensia effectum ut fastae essent, uti rustici, qui nundinandi causa in urbem veniebant, lites componerent.*

[70] Cf. the definition of *feriae* in Macrob. *Sat.* 1.16.5–6: *Feriarum autem publicarum genera sunt quattuor. aut enim stativae sunt aut conceptivae aut imperativae aut nundinae. (6) ... nundinae sunt paganorum itemque rusticorum, quibus conveniunt negotiis propriis vel mercibus provisuri.*

[71] Each of the Tubilustrium *feriae* on 21 April, 23 July and 21 August, is followed by an *N* day, on the third occasion by an *EN* day. The Divalia, however, with the *C* day on 22 December, do not fit this pattern, which approximates to the Tubilustrium-*QRCF* complex.

[72] Macrob. *Sat.* 1.15.21.

of the ordinary citizen: at the level of the private house cult, the orientation days were, without exception, given the same status as feast days (*dies festi*), and were all celebrated by dressing the hearth and offering a prayer to the Lares.[73] Mere recollection of what the *nundinae* had replaced was thus insufficient to provide a clear resolution. To compound the problem, the new *nundinae*, by their moveable nature, had the propensity to 'collide' with established *feriae*. What was to be done?

The solution may have resided in a 'softer' formulation of the day's character, in the manner of a *dies religiosus*; in relation to the procedure for people's assemblies, this corresponded to the practice for *feriae*, but with a relatively free field left for legal business. The concept was given expression in the term *dies sollemnes*[74] (obviously the result of an exercise in pontifical semantics), thus integrating the new, profane element in Flavius' calendar, the *nundinae*, with the sacrifice in the Regia: not only ritually, but also terminologically. In terms of semantics, the *dies sollemnes* are located between *sacer* and *religiosus*. As *religiosus*, they express an obligation, but a positive one, tending towards *sacer*: an obligation in respect of religious performance, not avoidance; in this, they approximate to *feriae*.[75] Characteristic of the term (and of the situation it relates to) is that the grounds for the obligation cannot be clearly asserted.[76] Flavius' codification allowed for no such flexibility: in the duality of *fas* and *nefas*, all or nothing, he opted for religious and political restriction, settling for 'nothing', that is to say the *nefas* label.

There was no objection on the pontifical side, and scarcely any elsewhere in the leading elite. Protest arose among those who were affected, the country people who regulated their lives on the basis of the nundinal cycle, and it was not until the disturbances of *c*.287 BCE[77] that a solution was conceded. The Livian tradition (as abridged by the epitomists), agrees with Granius Licinianus in emphasizing as the leading causes of those disturbances private legal problems connected with debts and the trials resulting from debts. According to Granius, one of the most important demands was that the *nundinae*, days when the population was naturally present in the city, should be made *dies fasti*, and so available for legal affairs.

One small problem stood in the way of this demand. Qualification of the *nundinae* as *fas* also made them available for holding decision-making assemblies. The solution was to bring a religious technical term, *dies comitiales*, into the (political) calendar. These days had nothing to do with *fas* or *nefas*, but belonged in the

[73] Cato, *Agr.* 143.2.

[74] Titus, *De feriis* ap. Macrob. *Sat.* 1.16.28.

[75] See e.g. Cato, *Orat.* 59, Sblendorio Cugusi; Suet. *Aug.* 75: *Festos et sollemnes dies profusissime, nonnumquam tantum iocalariter celebrabat.*

[76] See e.g. Liv. 1.31.4; Plin. *HN* 28.135; see also Liv. 3.36.2: *idus ... sollemnes.*

[77] Liv. *Per.* 11.11.

realm of propitious and unpropitious days. Formally speaking, they corresponded to the *dies proeliares*, days on which ultimatums or other provocations could be given to the enemy.[78] Neither term could be said to belong in the narrower sense to sacral law,[79] but, in practice, they filled the bill. Just as only *dies proeliares* were to be used for dealings preparative for war, only *dies comitiales* were *really* appropriate days for holding *comitia*. This rule was now used to arrive at a more precise definition of *dies fasti*. Entirely plausibly in the political circumstances, the juridical element was distinguished as being at the core of the definition of a *dies fastus*. The *dies fasti* – and only the *dies fasti*[80] – required in addition to be qualified as *dies comitiales*; as in a manner of speaking the most momentous element, this was included in the *fasti*, and the *F* label suppressed.

Before advancing a stage further in our historical reconstruction, we should satisfy ourselves once more of the basis of the argument, the question of the character of the *nundinae* prior to the *lex Hortensia*. In the absence of any indications to the contrary, the cited account by Granius Licinianus, passed down to us by Macrobius, to the effect that the law stipulated, in respect of the *nundinae*, '*ut fastae essent*',[81] can only be understood to mean that the *nundinae* changed their character. In fact, the generally accepted interpretation up to now has been that assembly days became court days, thus entailing a restriction of the plebeian element.[82] Now that it has been demonstrated that the category of *dies comitiales* was not previously available, this interpretation falls by the wayside: the only alternative prior to this period was *nefas*.[83]

The situation in respect of Kalends and Nones is more difficult. Here too, the sacral complexion of these days suggests that people's assemblies were excluded early on, and that Flavius accordingly categorized them as *N* days. Their characterization as *fas* by the *lex Hortensia*, a definition that at this point in time already excluded *comitia*, should be seen as a ruling by analogy.

But to return now to the final and most intriguing aspect of the *lex Hortensia*, which is in fact known for a quite different reason: its placing of *concilia plebis* on an equal footing with *comitia*. The law stipulated that plebiscites should have the status of general people's assemblies.[84]

The facts of the matter leave no room for doubt. The law's stipulation to this effect is sufficiently well attested.[85] The question that remains is, how the law fitted

[78] Macrob. *Sat.* 1.16.15–17.

[79] See Varro ap. Macrob. *Sat.* 1.16.18–19.

[80] MICHELS 1967: 107.

[81] Macrob. *Sat.* 1.16.30.

[82] BLEICKEN 1955: 23; HÖLKESKAMP 1987: 169; HÖLKESKAMP 1988a.

[83] Thus already UNGER 1895: 501. Cf. Gell. 20.1.47 (either anachronistic or referring to the monthly structure). Macrob. *Sat.* 1.16.34; Dion. Hal. 7.58.3 are later imaginings.

[84] HEUSS 1976: 33. See also BLEICKEN 1955: 12–18; HÖLKESKAMP 1988a: 292–301.

[85] Plin. *HN* 16.37; Gai. *Inst.* 1.3; earlier Laelius Felix *ap.* Gell. 15.27.4: *In eodem Laeli Felicis libro haec scripta sunt: Is qui non universum populum, sed partem aliquam adesse iubet, non 'comitia', sed 'concilium' edicere debet. tribuni autem*

into the political situation, and how it related to the calendar reform. *Communis opinio* in this regard is easily summarized: the only possible background against which the law can be understood is that the process of development of the nobility, that is to say the melding of patricians and the upper stratum of plebeians into a leading elite, must have been so far advanced that the people's tribunes, under whose direction *plebiscita* were called, were not actors in a 'class struggle', but organs of the balance of power within the nobility.[86] Nevertheless – and here we come to the relationship between the law's two components – the concession to the *plebs* is offset by an overall reduction in the number of assembly days.[87] Can the contradictions be resolved?

How if we were simply to reverse our evaluation of the two provisions? Economic and legal problems stood at the centre of the disturbances,[88] and a central element of the solution to those problems was seen to be a return to the 'natural' day for the hearing of cases.

Declaration of the *nundinae* as *dies fasti* on which no *comitia* could be held preassumes the designation of *dies comitiales*. But what of *concilia*, and *concilia plebis*? There would have been no desire at all to extend the prohibition to them, as these very institutions had gained greatly in significance in the previous decades. The solution found was that, if the *concilia* were to be given equal status with the *comitia* – and, in view of the circumstances, this can only have meant making them equally binding on the community – then their timing must be made subject to the same rules as applied to the *comitia*. Later on, it would be the people's tribune Clodius who, by means of a far-reaching bundle of laws, would seek to acquire the right to hold legislative assemblies, in this case *concilia plebis*, on *dies fasti* too.[89] The very idea was made possible by the grounds set out here: the fact that the people's tribunate occupied such a thoroughly integrated position.[90] This interpretation explains why, subsequently, no distinction was made in respect of the meeting times of general assemblies on the one hand, and *plebiscita* on the other. It allows us to discern the *lex Hortensia* as a whole, and as less 'surprising'. It demonstrates the central significance of the regulation of political processes in relation to time. And it shows yet again the care that must be taken when investigating the relationship between religion, the economy, and politics in respect of the organization of time, and the way that relationship is expressed in the '*fasti*'.

neque advocant patricios neque ad eos referre ulla de re possunt. ita ne 'leges' quidem proprie, sed 'plebisscita' appellantur, quae tribunis plebis ferentibus accepta sunt, quibus rogationibus ante patricii non tenebantur, donec Q. Hortensius dictator eam legem tulit, ut eo iure, quod plebs statuisset, omnes Quirites tenerentur.

86 Cf. HÖLKESKAMP 1988a.

87 An alternative view in MICHELS 1967: 106.

88 BLEICKEN (1955: 18–21) too sees economic problems as the main driving force.

89 Cic. *Sest.* 33.

90 See THOMMEN 1989.

5.6 Implications for the Historiography of Roman Religion

Nobody has given more acute expression to the significance of the *fasti* for research into Roman religion under the Republic than GEORG WISSOWA, author of what is still the definitive study of the history of Roman religion:

'By far our most important source for the knowledge of ancient Roman religion, as it was prior to the influence of Greek models, is the Roman *Calendar of festivals*, whose original state we are able to reconstruct with complete confidence on the basis of specimens from the First Imperial Period,[91] considerable fragments of which are available to us.' We are greatly indebted to THEODOR MOMMSEN for having shown (*CIL* I^1 pp. 361ff. = I^2 pp. 238ff.) that, in the details inscribed in large letters, 'we ... possess the oldest record of the Roman calendar, in the form that the Romans themselves regarded as reflecting the year's structure as set out by Numa, and which endured without any alteration for the entire period of the Republic until C. Julius Caesar'.[92]

Then, at the beginning of his section on the individual gods, he explains the importance of this fact:

> But our knowledge of them [the gods] would be scant in the extreme, if we did not, in the oldest table of Roman festivals, preserved for us by the stone calendars (above, p. 2f.), have an authentic document that grants us a virtually complete survey of the circle of gods worshipped by the Roman populace, at a period of its existence that is the earliest discoverable by the methods of historical research. By means of this document, we know the cycle of constantly recurring feast days belonging to the oldest religious system, and, as a large portion of these *feriae* reveal by their very names their attachment to the cult of particular divinities, while, for others, the same is assured by irreproachable witnesses, we are given a long series of names of gods from the earliest period, while, at the same time, the position and sequence of festivals in the year often inform us of the significance of the divinities in question, and of the associations that existed between many of the cults.[93]

There would have been no need for this long quotation if it were not for the fact that the passage embodies the *communis opinio* that has formed the basis for all studies of the history of religion under the Republic to this day. The calendar is judged to be a religious document of central importance. Our investigation of Flavius' publication of the *fasti* deals the above interpretation a severe blow. To reiterate the most important outcomes of that investigation:

91 There was no knowledge of the Fasti Antiates maiores before WISSOWA 1923a.

92 WISSOWA 1912: 2–3.

93 WISSOWA 1912: 19–20.

1 If the feast-names reflect the terminology of the end of the fourth century, and perhaps even to some extent have their origins in this period, it is not appropriate to use them in a historical analysis of the strata of early Roman religion: they arose in response to the need to characterize widely differing *feriae*, using one word for each. There was thus no room for epithets in the list of names. If, instead of a god's name, the basis used is a ritual component, this signifies nothing more than that, shortly before the year 300, this element was seen as being the more easily remembered. To conclude the existence of a pre-theistic cult level, as DEUBNER and many after him have done,[94] is impermissible. There may never have been any doubt that, for example, the Equirria were associated with Mars.

2 The more important a divinity was, and the more *feriae* it possessed, the slighter the possibility of finding repeated instances of it in the *fasti*. One entry named 'Junonalia' was feasible; used more than once in the calendar, the name would have ceased to be instructive. The same applies to Jupiter and Mars, and perhaps Minerva.

3 The insertion of intermediate days not having the character of *feriae* typifies the calendar so thoroughly that the pontifical origin of this rule can scarcely be disputed. But how was the situation at the cult level below the *feriae* threshold, i.e. the level that finds no adequate portrayal in the calendar? This question becomes interesting in respect of the festivals that occur on more than one occasion: the Carmentalia on 11 and 15 January, the Lemuria on 9, 11, and 13 May, the Lucaria on 19 and 21 July: were these festivals perhaps seen in sacral law as connected, three- or five-day cult sequences? Were there shifts of emphasis, giving the following day a different cult character that Flavius did not see as significant?[95] But the question can be posed in a different, more difficult form: is it possible that festivals with different names were much more closely associated than is immediately apparent from Flavius' calendar?[96]

The answer to many questions must remain open, at least in the context of this study. Perhaps the most important finding of the present chapter is that the status of the *fasti* must be fundamentally redefined. What we have before us is not a religious document extracted from the pontifical archive, and capable of being seen without further ado as a record of pontifical (or even Numan) theology and structure. 'Numa's cycle of festivals' should not simply be seen as a system that passed down from the period of the early monarchy into the Middle Republic: and, in terms of its content, it never existed as an entity under sacral law. Behind the abbreviations of Flavius' calendar lie religious dates, but these appear only in

[94] See e.g. WISSOWA 1902: 26 (= 1912: 29); DEUBNER 1911: 331–2; GJERSTAD 1961: 211.

[95] In the case of the Carmentae, the double name Porrima/Postvorta gives rise to speculation; see KÖVES-ZULAUF 1990: 272–4.

[96] See discussion in RÜPKE 1997.

a heavily refracted form, through a prism that entirely absorbs as much as it distorts. Flavius' politico-legal, graphic calendar cannot form the basis for a religious-historical interpretation; for this, we must first use Flavius' publication and other sources to reconstruct the calendar of the religious 'subsystem'. For the moment, however, we shall confine ourselves to externals.

5.7 Variants on Stone and Paper

Nefas piaculum gave us the key to the character and political intention of Flavius' calendar. Our newly won resolution of this most important and at the same time most baffling calendar abbreviation may now illuminate some smaller problems that belong, not chronologically but materially, to the same context. The first involves a justification of antiquarian literature, which, when read in the manner I have suggested, does not ignore the *P* element in its explanation of the *fasti*; on the contrary, it brings it strongly to the fore. All our sources, when speaking of *dies fasti* and *nefasti*, address the *piaculum* rule. What the antiquarian writers neglect to do is to explain the difference between *N* and *NP*: a difference that provided us with the key for deciphering the project of Flavius and Claudius, and a difference the inventors of the *fasti* were least concerned to discuss.

Like the abbreviations for *dies fissi* (*QRCF*, *QStDF*), *NP* and *EN* too teach us that *N*, *F*, and *C* in the *fasti* should not be resolved as *dies nefasti/fasti/comitiales*, but as *nefas*, *fas*, and *comitia*: as already suggested by WISSOWA,[97] an appropriate correction should be made in all editions of the calendar. The researcher's first task, when considering instances of literary and epigraphic usage, is to investigate their mutual identification; in the case of the calendar at least, this must never be taken for granted.

We are also now able to resolve the most obstinate, niggling problems remaining in some Imperial-Period *fasti*: their authors too – in common with any of their contemporaries who were at all well-informed – knew the correct meaning of *NP*. When, in the Fasti Venusini and the Fasti Pinciani, only *N* is always written instead of *NP*, the use of the shortened form is very understandable: for every non-magistrate, the *P* embodied only the hidden signification *feriae*, and this information was transparently obvious to users of both calendars. The decision of the author of the Fasti Tusculani not to give character letters to feast days already written out in full, when such additional information would have been redundant in most cases, may have been due to similar considerations. Finally, the *NF* in the Fasti Pighiani is now resolved: it simply stands for *N(efas) F(eriae)*, rendering the official *P* as the more 'user-friendly' *feriae*. This solution is placed beyond any doubt by the entry for the Volturnalia, which should have *NP*, but has only *N*; here, a

[97] WISSOWA 1909: 2015, 1912: 438 n. 1. DEGRASSI 1963: 331 disagrees, pointing to *endotercisus*.

long form of *NF* is actually used, the *Fer* taking the place of the divinity's name in the dative, which in other cases directly follows the abbreviated form. Accordingly, the *N. Fer* on 24 September[98] should also not be regarded as an error:[99] it precisely renders *NP*, which has already been demanded by the previous day, and is in fact borne by it in the form *NF*.

But to return to the chronology: we have arrived at the third century, and yet the calendar is still not complete. To summarize: it comprises the nundinal letters, the orientation days, the daily characters (from about 287 BCE also *C*), and the familiar, explanatory festal abbreviations in 'large' letters. A half bureaucratic, half pontifical document, it reveals little about cults, names, and judgements under sacral law, still less about 'Numa's cycle of festivals': no entry is necessarily older than the end of the fourth century. Entries in small letters are entirely absent. Before we can pursue the development of the calendar's written form up to the point when it corresponds to the Fasti Antiates maiores of the first century BCE – our starting point, as the oldest extant specimen – we should, however, attend to an event that may have had little to do with the look of the calendar, but much to do with the way it was used.

98 *Inscr. It.* 13.2.219.

99 Thus DEGRASSI 1963: 514.

6

The *Lex Acilia* and the Problem of Pontifical Intercalation

6.1 The Nature of the Measures

One of the most clearly attested elements of the history of the Republican calendar is the existence of a law that governed intercalation in 191 BCE, and was associated with the name of Manlius Acilius (Glabrio), a consul for that year.[1] Our source fails to go into the detail of the stipulations of this law; they cannot have encompassed the introduction of intercalation, whose existence is attested prior to this time. There can, however, be no doubt as to the aims of the measure. By the beginning of the second century BCE, the discrepancy between the natural year and the civil year had reached such proportions as to make it appear desirable to revise the rules governing the intercalation process. An account by Livy of an eclipse of the sun that occurred on 11 July 190 allows us to establish that, owing to repeated failure to implement intercalations, the calendar was by then nearly four months in advance of the solar year.[2]

What can have been the nature of these changes devised at the beginning of the second century? A survey of sources for the intercalation process[3] quickly reveals that antiquarians often distorted the historical narrative by the inclusion of elaborate Greek theories on the subject, which is all the more reason why we should pay heed to the account of a specifically Roman development: the transfer of control over intercalation to the *pontifices* at a particular point in time.[4] Although

[1] Macrob. *Sat.* 1.13.21: *Fulvius autem id egisse M'. Acilium consulem dicit ab urbe condita anno quingentesimo sexagesimo secundo, inito mox bello Aetolico.*

[2] Liv. 37.4.4; in fact 14 March (Julian).

[3] Michels 1967: 146–60.

[4] Censorinus, *DN* 20.6–7; Solinus. 1.43; cf. Amm. Marc. 26.1.12.

The Roman Calendar from Numa to Constantine: Time, History and the Fasti, First Edition. Jörg Rüpke.

it is immediately tempting to conclude that the account in question is associated with the *lex Acilia*, this widely held hypothesis[5] lacks firm proof,[6] and has been vehemently disputed in recent years. It is said that, owing to the nature of the entire calendar project, the *pontifices* must have been entrusted with intercalation from the beginning,[7] that there is nothing to suggest that this task was transferred at the beginning of the second century, and that the only necessary – and demonstrable – effect of the law was an acceleration of the intercalation process, whose effect must have been to reduce the discrepancy between the calendar and the solar year to a tolerable measure.[8]

The third point is the least difficult to address. Research into intercalations in the first third of the second century shows that, after an increase in the 190s, at the beginning of the 160s the difference had reduced to two and a half months, and this only by dint of relatively frequent intercalation.[9] Precise reconstruction of individual intercalation events remains irrelevant to the question of concern to us here, as, evidently, the *lex* produced no measure that removed the discrepancy entirely or largely, let alone permanently, within a short space of time.[10] This does not allow us to draw any conclusions as to the role played in the law by the *pontifices*; but it does prove that the measure arrived at was general in nature, and did not amount to an improved system of intercalation.

6.2 The Ritually Correct Method of Intercalation

The three elements at the core of Flavius' new structure were the Republican system of intercalation, the abolition of true lunar months, and the *nundinae*. The technical details have already been adequately described; but two important aspects remain to be addressed: how was the decision in respect of intercalation arrived at, and how announced; and why did the intercalary month begin on the day after the Terminalia of 23 February? Would the adjustment not have been much more easily made at the end of the month? It appears that this decision too was one of the 'compromises' between religious and politico-economic

[5] Important SOLTAU 1889: 32, 213.

[6] MICHELS 1967: 102.

[7] PAIS 1909: 201–11.

[8] WARRIOR 1992.

[9] The eclipse of the moon before the battle of Pydna, which Livy places by contemporary reckoning on 3 September 168 BCE (44.37.8), occurred on 21 June (Julian); see WARRIOR 1991: 80.

[10] The systems postulated by both DEROW (1973: three successive intercalations, then a two-year cycle) and WARRIOR (1992: successive intercalations to achieve parity, but with exceptions) fail to do justice either to the astronomical knowledge of the period or to the political demands. Notable studies of the chronology of these years, after CAVAIGNAC 1915, are: BELOCH 1918, 1929; MELONI 1954; MARCHETTI 1973; RAEPSAET-CHARLIER AND DEMAN 1974; DEROW 1976; BRIND'AMOUR 1983 (with further literature); see also BRISCOE 1991. For a critique of individual points in WARRIOR's argument see below.

requirements arrived at in connection with the reform, and so could help us to find an answer to the first question, which belongs in the same area.

None of our sources describes the announcement of an intercalary month. Cicero's correspondence in 51 and 50 suggests that there was no question of any standard practice at that time, and that everyone attempted to influence the decision in his own favour.[11] The campaign against intercalation in the year 50, which would have lengthened Cicero's period of office (in his opinion unnecessarily) in extending Caesar's proconsulate by just one month, began as early as the middle of June in the previous year; foremost among the supporters of this campaign was Q. Hortensius L. f. Hortalus, Cicero's respected colleague in the college of augurs.[12] But the decision was taken in the pontifical college, where C. Scribonius C. f. C. n. Curio, *pontifex* since 51, was unable to prevail in Caesar's favour.[13] We do not know when the decision was published. Cicero is still ignorant of the result in letters of 13 and 20 February; it does not appear likely that M. Caelius' letter telling him of the positive outcome was written before the end of February of the year 50.[14]

WARRIOR reaches a similar conclusion for the first decade of the second century BCE: very late announcement of intercalation appears to have been traditional practice.[15] But it remains unclear who made such announcements. That the decision during the Late Republic rested with the pontificate is beyond dispute: it is attested both by Cicero's correspondence and by the Imperial-Period antiquarian sources cited at the beginning of this chapter. We cannot, however, conclude from this that publication of the decision occurred in an institutional context.

In respect of an earlier period, the question remains as to who made the decision implemented by the ritual agents of the pontifical college. As regards the situation at the time of the reform, the quite simple answer might be no-one. The sole justification for the complexity of the system involved is that it was intended to be automatic. Just as anyone could know in advance whether the Nones would fall on the fifth or the seventh day of the month, they could also know (soon: should have been able to know) whether intercalation would occur, and the extent of any such intercalation. Every second year, twenty-two days were added to February. The average length of year aimed for by this means was 366 days. This total was increased to 366.25 by the addition of a further day every fourth year (thus every second intercalary year). The only mathematical error in comparison

11 WARRIOR 1992: 128.

12 Cic. *Att.* 5.9.2: *memento curare per te et per omnis nostros, in primis per Hortensium, ut annus noster maneat suo statu, ne quid novi decernatur. hoc tibi ita mando ut dubitem an etiam te rogem ut pugnes ne intercaletur. sed non audeo tibi omnia onera imponere; annum quidem utique teneto.*

13 Cael. *Cic. fam.* 8.6.5; Dio Cass. 40.62.1–2. For a contrary view MALITZ 1987: 108.

14 SHACKLETON BAILEY 1977: 1.415.: cf. Cic. *Att.* 5.21.14: *Cum scies Romae intercalatum sit necne, velim ad me scribas certum quo die mysteria futura sint* (13 February; Cic. *Att.* 6.1.12: *Quinto togam puram Liberalibus cogitabam dare (mandavit enim pater). ea sic observabo quasi intercalatum non sit* (20 February). See also Cic. *Fam.* 7.2.4. SHACKLETON BAILEY 1977: 1.351.

15 WARRIOR 1991: 84 n. 29, 1992: 129, 144.

with the Julian year lay not in the missing quarter of a day (which, in the case of Egypt, persisted after the Julian reform), but in the length of the basic year,[16] at 355 days exceeding the 354 of the conventional lunar year. This error is hard to explain. Unless we choose to assume that our information is wrong, then the key may lie in the desire to avoid further months with even numbers of days, and in the impossibility of making February shorter still.

In theory, the civil year should subsequently have fallen further behind the natural year by one day each year. This error could have been corrected by omitting the occasional intercalation. We do not know whether this happened; but, by the end of the third century, the remedy, if it had ever been such, had turned into its opposite: frequent failure to intercalate allowed the civil calendar to hurry ever futher ahead of the solar year. The automatic element had fallen by the wayside: by whose responsibility we do not know.

Our final attempt to pursue this question further must be founded on clues provided by the intercalation apparatus itself, beginning with the curious choice of starting point, and the shape of the final portion of the month. Again, it appears appropriate, in seeking a solution, to proceed initially by complicating rather than simplifying the problem, and to ask first how intercalation functioned prior to the reform.

In the lunisolar system that preceded the Republican calendar, the intercalary months too were true lunar months. As the Flavian year began, technically speaking (nundinal letter *A*!), on 1 January, and intercalation at the year's, or at least the month's end represents the simplest solution, we may conclude from the positioning of the intercalary month at the end of February that such a location had its origin in the pre-Republican system. At that early period too, therefore, the thirteenth lunar month was inserted, not appended.

In such a system, the criteria for intercalation must be sought in the need to synchronize with prominent events in nature. Less strictly structured 'calendars' respond to the same need, merely in identifying benchmarks for the different seasons of the year. The means used may be a 'bloom calendar', as in the Andaman Islands, or, in the Trobriand Islands, the anticipated emergence of a particular sea-worm; but less exotic organic phenomena or astronomical observations may also be called upon.[17] The equinoxes and solstices are widely favoured, the latter being easier to observe than the former.[18]

At the latitude of Rome, the most convenient criterion was the winter solstice, which is perhaps addressed in the calendar in the form of the festival of Divalia. The 'civil' calendar's annual advance in relation to the natural date of the winter solstice (by eleven days, although the observer did not need to know this) was determined at this time, and intercalation was implemented as soon as the

16 Not seen by SAMUEL 1972: 159.

17 See MALITZ 1968, especially 91.

18 SAMUEL 1972: 17 and 23 n. 1.

difference approached the amount of a whole month. At the time of year when this assessment was being made in Rome, the calendar date was already approaching the end of January, so that intercalation was out of the question before the February Kalends.

There is no need here to repeat our discussion of February's location at the year's end. The month's cult structure included a substantial role for Juno; this began with the Kalends, subsequently devoted to Lanuvine Juno Sospita or Juno Mater Regina, and extended up to the Lupercalia.[19] The second half of the month was the more important. At midday on the Ides, at the onset of the second, 'waning' half of the February Ides, there began a *nundinum* called *dies parentales*.[20] It was devoted to the private cult of the dead, and closed on 21 February with the Feralia, followed by the Caristia (22 February), a family festival celebrated outside the framework of the cult of the dead, and devoted to the reconciliation of relatives.[21]

When, for example in inscriptions, a single *dies Parentaliorum* is mentioned, it may be assumed that the final day, the Feralia, is meant. Ovid's description, though, clearly indicates that an entire period was combined under a single set of rules.[22] The most sweeping of these was one forbidding marriage during the period, and was also associated with the feast days of the dead at the Lemuria.[23] Similar provisions applied on days of *mundus patet*, when contact was established with the underworld by the uncovering of a certain pit.[24]

Despite the fact that restrictions on public life were usual on such days, two stipulations for the *dies parentales* were without parallel: first, Lydus tells us that magistrates had to behave like private citizens until 22 February;[25] and, second, Ovid reports in the passage already cited that the temples were closed, and that every cult in them fell into abeyance. When WISSOWA and LATTE extend the rule on closure to the Lemuria at the beginning of May, they are certainly following a hint in Ovid, but their reading of it is too superficial by far.[26] Ovid describes the closure in May as a practice of Rome's foundation period, which is the setting for the episode he relates for the day; and he takes the Feralia as the starting point for his analogy.[27] The entire necessity of Ovid's hypothesis, however, lies in a historical construct that we are not required to imitate: that, as February's (and January's) feasts could not, of course, be celebrated in Romulus' ten-month year, the May

[19] Paul. Fest. 75.23–76.5 L; cf. Varro, *Ling.* 6.13. See WISSOWA 1912: 185, 188.

[20] *Inscr. It.* 13.2.241 (Fast. Filoc. 13 February) – *Virgo Vesta(lis) parentat* (see PORTE 1989: 75) – and Lydus, *Mens.* 4.29; SABBATUCCI 1988: 47–8.

[21] Ov. *Fast.* 2.617–38.

[22] Ov. *Fast.* 2.557–68. See RÜPKE 1994a.

[23] Ov. *Fast.* 5.485–92.

[24] See Festus, *Gloss. Lat.* 144.14–146.2 L and Macrob. *Sat.* 1.16.16–18 (Varro).

[25] Lydus, *Mens.* 4.29.

[26] WISSOWA 1912: 235; LATTE 1960: 99.

[27] Ov. *Fast.* 5.485–6: *Fana tamen veteres illis clausere diebus, / ut nunc ferali tempore operta vides.*

festival of the dead served as a surrogate for or precursor of the Feralia / Parentalia, and was subject to the same ritual structure.

Two festivals marked *NP* are located within the *nundinum* of the Parentalia: the so-called Lupercalia on 15 February, and the Quirinalia on the 17th; thus contradicting the notion that the cult was dormant during this period. If we restrict the scope of Ovid's account to 'temple cults', the first-named festival no longer constitutes a problem: the rites performed on the day were not associated with a temple; if the *lupercal* represented an architectural entity at all,[28] it may not have fallen into the category of either an *aedes* or a *templum*. In accord with Lydus' account is the circumstance that state authorities were not present in their official function; the pseudo-gentilician priesthoods of the Luperci were virtually or completely naked. It is no longer possible to define the cult function of the *flamen Dialis*. Ovid alone notes that he was formerly responsible for the cult of Pan, or Faunus.[29] As the flaminate remained vacant for decades, until it was occupied again in 14 BCE by Ser. Cornelius Lentulus Maluginensis, Ovid appears to assume that Augustus' refusal to take part in the festival – perhaps owing to its ambivalent reputation – was not innovative. There is no way of clarifying whether this view is correct, or whether Ovid is merely referring to some other association of priests with the cult of Faunus.[30] Of significance for the general cult situation at the period, in any event, is the fact that the temple of Faunus *in insula*, dedicated at the beginning of the second century BCE, celebrated its foundation day on 13, not 15 February: the corresponding ritual, along with the *sacra idulia*, could have ended before midday on the Ides.

More of a problem is the second date, the Quirinalia in the middle of the *nundinum*. Wherever these *feriae* are discussed, reference is made to their designation as *feriae stultorum*: whoever could not participate in the Fornacalia festival, which was celebrated on various days in the individual *curiae*, or who could not assign himself to a *curia*, caught up with the cult on this day.[31] Our sources cannot say what the Quirinus cult comprised on this occasion. A proven cult centre was dedicated as late as 293 BCE, by L. Papirius Cursor.[32] Thus the theory that this cult 'faded away' is unconvincing: construction of the temple at such a late period entitles us to assume an extant cult context. In seeking clarification of its nature, we should not overlook the information that is there for all to see: *feriae stultorum*, the final day of the Fornacalia, and Quirinus, converge in the concept of the *quirites*.

While the etymological association of the adjectival form *quirinus* with the 'men's clubs' of the *cūriae*, derived from **coviria*, has proved problematic,[33] the connection with *quirites* is irrefutable. This concept does not signify the members

28 ULF 1982: 29–38.

29 Ov. *Fast.* 2.281–2.

30 ULF 1982: 54–8; cf. KÖVES-ZULAUF 1990: 286–9.

31 Varro *Ling.* 6.13; Ov. *Fast.* 2.513–32; Festus, *Gloss. Lat.* 304.5–11, 418.33–420.6 L; Plut. *Quaest. Rom.* 89.

32 Liv. 10.46.7.

33 See KOCH 1953: 9–14; ERNOUT AND MEILLET 559; RADKE 1981: 284–6.

of the *curiae*, as the problematic etymology would suggest, but, to follow the unambiguous semasiology, embraces all Roman citizens, defined by political criteria.[34] The antiquity of the god Quirinus is a question we need not enter into here. His feast-name and temple tend to locate him at the end of the fourth century. Neither need we look for missing cult activities: the *feriae stultorum*, as civic culmination and transformation of the Fornacalia, whose theme was the roasting of the primitive emmer (*far*), provide a sufficient basis for the existence of god and festival. It thus appears that the Quirinalia, while, as a festival, probably possessing a central cult component in the shape of the temple's dedication day (in which the *flamen Quirinalis* was perhaps involved), and therefore being qualified *NP*, primarily served as a cult date for *quirites ob Fornacalia feriati*, inserted into the subdued context of the *nundinum* dedicated to the cult of the dead.

It was this complex relationship between private and public cult that determined the designation of the closing day, the Feralia. By definition, the Republican cult of the dead was not a public, but a private cult. There was, therefore, no need to bar the day to the catalogue of public activities that had a religious relevance, and mark it *N(efas)* from a legal standpoint. As, however, the responsibility for providing for one's own dead was seen as *conditio humana*, making *de facto* everyone *feriati*, the warning *Piaculum!* contained in the note *FP* was entirely justified: one more example of the possible discrepancy between the pontifical assessment of a day on religious grounds, and the political and juridical view.

After the purely private festival of Caristia, the Terminalia at the end of the period again assumed a public as well as a private character.[35] Ovid has it that the 'central' public cult act took place at the sixth milestone on the *via Laurentia*, presumably at the outer limit of the *ager Romanus*.[36] This, however, touches only one segment of the milestone cult, which was celebrated everywhere as a neighbourhood festival. It was in this guise that the Terminalia too took their place among the cult dates following the February Ides. In different rituals, the definition of social relationships was renewed: in communion with ancestors, regarded as integral members of the social system; in communion with the extended network of living relatives; and in communion with neighbours.

I have no intention here of giving a detailed interpretation of the Lupercalia; but there might be broad agreement in assigning a central role, in their mythical aetiology, to the 'social order' element of Rome's founding culture, and, in their ritual practice, to the element 'definition of sexual roles'. In the cult expression of inclusiveness – all must (be able to) celebrate Fornacalia – the Quirinalia had as their theme the social frontier represented by membership both of society as a whole, and – *ex negativo* – of the important social entity represented by the *curiae*.

[34] See e.g. PORTE 1981: 326–8; for an alternative view PALMER 1970: 157–8.

[35] Comprehensively: PICCALUGA 1974; see also PICCALUGA 1977: 47–52; GLADIGOW 1992.

[36] Ov. *Fast.* 2.679–84; see RÜPKE 1990a: 33–4.

With the passing of the *nundinum* of the Parentalia, the question arises as to whether the temples were reopened on the Caristia. If the association established by Ovid between the absence of the temple cult and the festival of the dead is not, or not quite, correct, it seems possible that the matter of the temples should be associated with the approaching year's end, and that the closure lasted until 'New Year'. There is no positive evidence for this. Nevertheless, we learn that on 1 March the door decoration of some important sacral structures, the Regia, and the official residences of the *flamines*, was renewed, and the fire in the Aedes Vestae newly kindled.[37] And counter-arguments? The Caristia took place in private, and thus required no temple. The Terminalia were celebrated in the open; the *terminus* in the Capitoline temple of Jupiter had an opening in the ceiling above it:[38] even when the temple was closed, it would accordingly have been under the 'open' sky on its feast day; nevertheless, its cult was allowed only in the open. There is no indication that a temple was required for the Regifugium; the Equirria on 27 February took place on the Field of Mars, thus even outside the *urbs*.[39] With the exception of the temple of Quirinus already discussed, we have no record of a dedication date for a temple in the second half of February. The first attested cult use of a temple occurs on 1 March: the *matronae* met in a temple of Juno.[40]

Again: were the temples closed from the February Ides until 1 March, perhaps for the entire intercalary month? In this form, the question is mistakenly posed. Roman temples were in any event normally closed;[41] it was not the act of closure, but that of opening, for example the opening of all temples in the context of a *supplicatio*, a day of supplication, that had a strong symbolic character. The occasional visitor had to pay the *aedituus*, the sacristan, an *obulus* if he wanted to view the artworks contained within.[42] This does not rule out exceptional use of the symbol of temple closure: in a Pisan inscription, closure of the temples features among honorific decrees for Augustus' grandson Gaius Caesar, at the head of measures decided upon for the funeral rites to mark the closure of Gaius' public life.[43] The customary actions associated with 1 March show that the symbolism of doors and new fire[44] became part of the rites for the change of year, but not in the form of total closure and reopening. Ovid's account should thus be interpreted in terms not of the fact of closure, but of an enormous reduction in the multifarious, many-stranded activities of the cult. This measure of reduction continued until the beginning of the new year, which saw a sudden renewal of activity at many locations, and in many cults. The 'closed' period was

37 Ov. *Fast.* 3.135–44.

38 Paul. Fest. 505.22–4 L; Ov. *Fast.* 2.669–72; Serv. *Aen.* 9.446. See also Dion. Hal. 3.69.5–6.

39 Ov. *Fast.* 2.860.

40 Ov. *Fast.* 3.205.

41 See e.g. Tac. *Ann.* 2.82.4 (night); 4.70.4; WISSOWA 1912: 476–7.

42 Cic. *Leg.* 2.25; Sen. *Ep.* 41.1; Tert. *Ad nat.* 1.10; *Apol.* 13.

43 *CIL* 11.1421 Z 22–3 = *ILS* 140.

44 Solin. 1.35.

characterized by a reduction in public activity, which did not exclude the occasional, exceptional ritual, such as a triumph.

To return to our progress through February, the Terminalia was followed by the Regifugium. It is not only its name and its position in the calendar that suggest an affinity between this date and the *QRCF* days: although both of the relevant passages in Festus' Lexicon, constituting by far our most comprehensive testimony, are badly damaged, it is possible to discern that Festus (that is, probably, Verrius Flaccus) established references between the two entries.[45] Before we turn our attention to detail, and to the connection with intercalation, two points may be registered straightaway. The first concerns the date. If our supposition of a connection with the *QRCF* days is correct, the disparity in calendar dates is striking. The constitutive feast day, the Terminalia, lies two days after the *'nundinae'*: not on 21 February, but on 23 February, the eleventh day after the Ides, counting inclusively. It is not hard to see that the reason for this lies in the *nundinum* of the festival of the dead and the triad of festivals concluding the phase. But the difficulty remains. The violation of the fundamental nundinal cycle appears to be calculated. With the impending year's end, the very structure of the calendar disintegrates.

The second point: the day's ritual content. Plutarch leaves no room for doubt that, as the name itself indicates, this concerned a ritual of flight, in which the *rex sacrorum* ran away after a sacrifice on the Comitium.[46] Such an event would have further accentuated the playing out of the cult crisis, after temple 'closure' and nundinal dislocation. The modifications, and the contrived disruption of the routine of important public rituals at the end of the year, prepared for the renewal on 1 March, and gave it its meaning. Some Greek states had comparable rituals of dissolution, the most well-known of which is perhaps the *apopompē* ritual on Lemnos.[47]

We should avoid forming an exaggerated idea of the social reality of this 'dissolution'. Only a few ritual symbols were affected, and they were dramatized in Rome through the medium of a single person. The ritual marked the dissolution of the nundinal structure by the flight of the individual who, in a special way, represented chronological order, with its varying but recurring elements, on the Kalends, the Nones, and the festivals that prefigured the Tubilustria-*QRCF* complex. The *rex sacrorum* disappears; the ritual category of space turns into that of time. But even this ritual dramatization must not be exaggerated. The date of 'New Year's Day' was certain: since the middle of the fifth century it had been there to be read on the tablets bearing the nundinal letters. The Equirria occurred between king's flight and New Year, bridging the period of 'disorder': held immediately before the new moon, they prepared the way for the reestablishment of order with the new month and year.

[45] See Festus, *Gloss. Lat.* 310.12–21 L and 346.22–36 L; Ov. *Fast.* 2.685–6; Auson. 14.16.13–14 Green (385 Souchay, p. 105.13–14 Prete).

[46] Plut. *Quaest. Rom.* 63.

[47] GRAF 1985: 236.

What effect did this scenario have on the problem of intercalation? As has already been established, there could be no question of announcing intercalation, which always inevitably affected an entire lunar month, before the February Kalends. Was this the moment for the intercalary month to be proclaimed? The answer is, no. The problem becomes clear if we give a thought to entirely formal factors. First, there is the question of the re-designation of the day. Even if it was not classified as *feriae*, its particular dedication to Juno would have to be altered after dawn had broken. This could not be expected of the goddess, the cult personnel prepared for her festival, or those attending. As ever, the difficulties are embodied by two differing standpoints: in organizational terms, it would be a question of finding a moment at which the announcement could be efficiently managed, technically speaking; in terms of sacral law, it was desirable to avoid retrospective re-designation of a day, or the interruption of ritual processes. The expression *inter*-calare itself, 'call in between', demonstrates contemporary awareness of the problem.

The next opportunity occurs on the February Nones, already the day when the *rex sacrorum* is due to announce the *feriae menstruae*.[48] The problem that now arises is, for what moment should the intercalary month be announced? Two considerable sacral factors speak against the end of the month: the context of the two Equirria of 27 February and 14 March would be lost, and the 'period of anarchy' at the end of the month would go on for too long. In order to keep the year's end running on safe cult lines, the task is thus to find a moment that, while causing no interruption to the accumulation of cult structures lasting until the Terminalia and beyond to the Regifugium, at the same time in some way precedes the Regifugium. Moreover, the chosen moment must be capable of accommodating one entire month in the midst of another.

A new difficulty for the *rex sacrorum* is, how to announce the *feriae menstruae*? Regifugium and Equirria surely both lie in February. This is where the true implications of interruption become evident. Intercalation, the addition of an entire month, will have to be structured in such a way that it nevertheless allows the festivals occurring at its end to be announced as February festivals. This is too complicated. The *rex sacrorum* announces the *feriae* as normal, and postpones the problem.

The next opportunity would normally occur at the *nundinae* after the Ides; but these are omitted on 21/22 February, in favour of the Feralia and Caristia. The assembly day (*QRCF*) that succeeds every *nundinae* is held two days late, in the wake of the Terminalia. This is the moment for the *rex sacrorum*: he sacrifices, and, instead of fleeing,[49] 'intercalates'. The next Kalends are 'additional' Kalends;[50]

[48] Thus already MOMMSEN 1858; HARTMANN 1882: 80.

[49] For an alternative view: SABBATUCCI 1988: 81. Sacrifice: Plut. *Quaest. Rom.* 63; see also Festus, *Gloss. Lat.* 346.22–36 L.

[50] Cato describes the intercalary month as *additicius (Dig.* 50.16.98).

a month is inserted. The day already begun is redefined; from *a(nte) d(iem) sextum kalendas Martias* it becomes *a. d. VI. Kal. Interkal.* The Early Republican dates for the second half of the month correspond with Flavian (and Julian) dates only in February; during this technically problematic period especially, there was evidently no desire to deviate from the old, two-*nundinum* interval between the Ides and the next Kalends. The redefinition functions smoothly: the day has no *feriae* character, and the main actor and the individual most affected is the *rex sacrorum* himself. The next month is now entirely routine; not even the *feriae menstruae* need to be announced. The announcement given on the February Nones remains valid: Regifugium will occur on *a. d. VI. Kal. Mart.* (as a routine date, it does not have to be announced as *feriae*), to be followed by the Equirria.

Intercalation in the Early Republican Calendar

(Italics: introduced days)

Kal. Febr. (first moonlight/narrowest crescent)

|

(variable: 3–6 days)

|

Nonae

|

Idus (full moon)

|

Nundinae after Idus ('Tubilustrium'): omitted

|

Feralia

Caristia

post d. XI. Id. Febr. = Terminalia

instead of Regifugium, intercalation = a. d. VI. Kal. Interc.

a. d. V. Kal. Interc.

|

pridie Kal. Interc.

Kal. Interc. (first moonlight/narrowest crescent)

|

(variable: 3–6 days)

|

Nonae

|

Idus (full moon)

|

post d. XI. Id. Interc. = a. d. VI. Kal. Mart. = Regifugium

|

a. d. III. Kal. Mart. = Equirria

|

Kal. Mart. (first moonlight/narrowest crescent)

6.3 Problems of Intercalation

The intercalation procedure as reconstructed here enabled two fictions: that the interpolated month was of nil extent, in both civil and religious law, counting as a single moment in time; and that it could be entirely assigned to February, that is to say including in particular the Regifugium and the Equirria.[51] The intercalated period began with the day of the now lapsed Regifugium, even though the *kalendae Interkalares* still lay in the future.[52] This interpretation also explains the later practice of effecting a twenty-three-day intercalation by inserting a single intercalary day on 24 February, prior to the intercalary month, rather than lengthening it by a day; the *dies bisextus* remained as an intercalary day in the Julian calendar.[53]

The religious aspects set out here also allow us to understand Macrobius' description of the intercalation as taking place *medio Terminaliorum*, 'in the middle of the Terminalia':[54] *Terminalia* encompasses the integrated structure (as explained at length above), comprising the 'real' Terminalia on 23 February along with the Regifugium following it; intercalation, which, according to sacral law, had no extent in time, happened at the centre of this ritual complex.

It is for this reason that the Regifugium's one-day advance in relation to the Ides in the decemviral *Interkalaris*, so that it now fell on the 'eleventh' instead of the 'twelfth' day after the Ides, did not entail the problem that might be suggested by a comparison with the Julian reform: Caesar added the ten new

[51] Q. Mucius P. f. P. n. Scaevola (*pontifex maximus* 89–82 BCE) in *Dig.* 50.16.98: *Cato putat mensem intercalarem additicium esse, omnesque eius dies pro momento temporis observat extremoque diei mensis Februarii adtribuit Quintus Mucius. mensis autem intercalaris constat ex diebus viginti octo.* The reconstruction of 27 days for the month's length in the *fasti Antiates maiores* is based on the surviving nundinal letters for the Ides and the Regifugium: of course, the 'leap' in nundinal letters occurs only once, with the *kalendae Interkalares*, while the transition from the final day to 1 March is seamless *(A-B)*.

[52] MICHELS 1967: 167–8. Liv. 43.11.13: *Hoc anno intercalatum est: tertio die post Terminalia kalendae intercalariae fuere.*

[53] See e.g. P. Iuventius Celsus in *Dig.* 50.16.98, pr.: *Cum bisextum kalendis est, nihil refert, utrum priore an posteriore die quis natus sit, et deinceps sextum kalendas eius natalis dies est: nam id biduum pro uno die habetur. sed posterior dies intercalatur, non prior: ideo quo anno intercalatum non est sexto kalendas natus, cum bisextum kalendis est, priorem diem natalem habet.* Ulp. *Dig.* 4.4.3.3; Censorinus, *DN* 20.10.

[54] Macrob. *Sat.* 1.13.19.

days to the ends of the months, so that, although the numerical dates, according to classical Roman practice, of all festivals in the second half of the twenty-nine-day months changed, their relative positions counting forwards from the Ides remained the same. But, if we interpret Terminalia and Regifugium as one integral cult complex, which always began on the 'eleventh' day after the Ides, the very same identity of interval also arises for the Regifugium, and the entire complex may then have borne only one name in the *rex sacrorum*'s announcement: a circumstance that Cn. Flavius, quite sensibly, did not wish to perpetuate in his written *fasti*. This would also explain the shortening of the second half of the intercalary month in relation to the normal double *nundinum*, and at the same time, in relation to February, its still further shortening to a span of only twenty-seven days in the decemviral calendar. We now see the converse effect of the interruption of such a compact ritual structure as the one comprising Terminalia and Regifugium: the intercalary month was thereby ritually 'quarantined'; normal time was safeguarded from 'non-time' by the festival period.

The intercalary month presented no problems for the Romans who were the recipients of the *rex sacrorum*'s 'unexpected' announcement. It did not bring with it any fixed sacral or legal obligations, and preparations for New Year's Day, a bare *nundinum* away, had surely not progressed so far that the 'countdown' could not be suspended. In a state the size of the *polis* of Rome prior to the close of the Latin Wars at least, this was feasible. The ritual event's well-calculated incalculability may have created a strong impression, and the king's role been given clear prominence by such a demonstration of control over time. It is easy to understand the concern, in post-monarchic Rome, to arrive sooner or later at a dispensation whereby, while retaining the ritual, the *rex sacrorum* might be deprived of his real power of disposal.

Flavius' reform of the intercalation system proceeded entirely on the lines of the pattern established for the lunar month. He only slightly reduced the lengths of the months, the required shortening of the intercalary month from its original twenty-nine or thirty days to only twenty-two (or twenty-three) being encompassed by removal of the period between Terminalia and *kalendae Interkalares*. This 'residue of February', which, at least colloquially and in terms of sacral law, must already have counted as part of the intercalary month, made up the 'disposable quantity'. It alternately disappeared entirely, or was reduced to one day immediately before the new Kalends.

Such a pedigree was not without consequences for the new system. This applies to the announcement of the 'automatic' intercalation, which was of course knowable in advance. Although the lengths of the months had been fixed, the date of the Nones continued to be proclaimed. We must, therefore, assume that the *rex sacrorum* continued to announce intercalations on the Regifugium: or rather, instead of enacting the Regifugium; and no longer on his own initiative, of course, but on a regular basis every second year. Note well, though: the *rex sacrorum*, not the *pontifices*;[55] and still on the date of the Regifugium. The continuity of the date was unproblematic, one might say immaterial, so long as the regular intercalary cycle was maintained. If it lapsed, though, and we know this in fact happened at some unknown time, that same continuity had fatal consequences: in the new system, the Regifugium did not lie just a 'week' before the Kalends of the intercalary month; it was identical with that day. In other words, the beginning of the intercalation, and with it the change of nundinal letter,[56] was not officially announced until after the event:[57] a practice that persisted until Caesar's reform. Seen in this light, Cicero's complaints and pleas acquire a new edge, and the degree to which the shaping of the chronological context of social activity had become an object of short-term political calculation and dissension can be seen, in a certain sense, as a measure for 'Republican decline'.

Intercalation in the Flavian Calendar

(Italics: introduced days)

Kal. Febr.

|

Nonae

|

Idus

|

[55] The sacral and (prosopographically underpinned) political position of the *rex sacrorum* in the context of the Roman priesthoods and the Republican system of governance in general requires new appraisal. The few sources that refer to him suggest that he also played a pivotal role in interactions with other members of the pontifical college; see e.g. Ov. *Fast.* 2.21: *Pontifices ab rege petunt et flamine lanas* (reported by Ovid in the context of his monthly reflections on the etymology of the month-name, the setting here is probably 15 February, *dies februus*, rather than the Nones, as postulated by PORTE 1989: 226). Cf. Serv. *Aen.* 10.228: *Virgines Vestae certa die ibant ad regem sacrorum et dicebant 'Vigilasne rex? vigila!'* (on the form of words, see also 8.3 and 2.148). The history of the senior pontificate from the third century BCE onwards nonetheless shows that the more 'political' functions of the collegiate priesthoods, and of the *pontifices* and augurs in particular, proved to be more fruitful avenues for the development of priestly institutions in the context of the diffusion of power within the elite that occurred under the Republic; against MOMIGLIANO (1969: 400), I assume only a gradual evolution in the significance of the *pontifex maximus*: although he may well already have had important sacral functions under the monarchy (see NORTH 1989: 612–13), he cannot be shown to have represented an autonomous centre of executive authority until Ti. Coruncanius. Taken together, these considerations make it appear entirely plausible that the *rex sacrorum* played a central role in the process of intercalation (SEGUIN 1988: 416, on the other hand, sees him as acting under the orders of the pontifical college, even in the matter of the monthly announcement of festivals).

[56] MICHELS 1967: 89 refers to this additional problem.

[57] Plutarch also tells of sudden, unexpected intercalations in the period before Caesar's reform (*Caes.* 59.2). But the advance already represented by the Republican calendar in comparison with the Greek is illustrated by the arbitrary way, in 304/3 BCE, the Greek calendar allowed the renaming of a month for the purposes of an initiation into the Eleusinian Mysteries, an event that was itself tied to the calendar (WOODHEAD 1989: 299–300).

Feralia
Caristia
Terminalia
instead of Regifugium, intercalation= K. Interc.
(in the 4th year: prid. K. Interc.)
|
Nonae
|
Idus
|
a. d. VI. Kal. Mart. = Regifugium
|
a. d. III. Kal. Mart. = Equirria
|
Kal. Mart.
=
=

Intercalation in the Julian Calendar

(Italics: introduced days)

Kal. Febr.
|
Nonae
|
Idus
|
Feralia
Caristia
Terminalia
a. d. bisextum Kal. Mart. (every 4th year)
Regifugium = a. d. VI. Kal. Mart.
a. d. V. Kal. Mart.
|
a. d. III. Kal. Mart. = Equirria
|
Kal. Mart.
=
=

6.4 Regulating Intercalation by Means of Laws

The main aim of Flavius' calendar reform was the establishment of months and years that were of regular length, and thus appropriate for the purposes of long-term planning. The aim was achieved by means of a calendar whose year remained parallel to the solar cycle, using a regulatory apparatus that was manageable and functioned without further intervention, while as far as possible retaining the substance of the old lunisolar system. We cannot tell, and it is in principle immaterial, whether this apparatus was safeguarded by a dedicated institution, beyond being recorded in graphic form. The new system's automatic character had not removed the roles of the individuals and institutions previously involved: the *rex sacrorum*, supported by the *pontifices*; both entities in fact continued to be concerned in the calendar's publication, although now deprived of any decision-making function.

We have no concrete evidence as to the manner in which the system came to be disrupted. Antiquarian sources point to the built-in error whereby, when functioning normally, the system produced an average year of 366.25 rather than 365.25 days.[58] Whatever measures were taken to mitigate the error, there was no longer any clear demarcation of rights and obligations. Although the *pontifices* were no longer automatically responsible for adjustments, at some point in the history of the Republic they were nevertheless entrusted with the regulation of intercalations. Was it the *lex Acilia* that provided this measure of institutional protection?

At the end of the Republic at any rate, no Roman politician appears even to have considered the idea of determining, either by resolve of the Senate or by a law *in comitiis*, that an intercalation should (or should not) occur in a particular year. This is not contradicted by the proceedings of the Julian calendar reform: Caesar – as dictator – decreed a new calendar, not a revision of intercalation under the old calendar (which would have sufficed). Misinterpretation of the intercalation mechanism by the *pontifices*, who intercalated *quarto quoque anno*, thus by Roman inclusive reckoning every three years, instead of *quarto quoque anno exacto* or *quinto quoque anno*, every fourth year, remained uncorrected until Augustus took the matter in hand, at a time when he himself was *pontifex maximus*.[59] On the other hand, laws concerning such matters as the renaming of months were also introduced and adopted at various times by non-pontiffs, instances being the *lex Antonia de mense Quintili* and the *lex Pacuvia de mense Sextili*, which created July and August.[60]

[58] Censorinus, *DN* 20.6; Macrob. *Sat.* 1.13.11.

[59] See Solin. 1.46–7; Macrob. *Sat.* 1.14.13–15. For an alternative explanation see HANNAH 2005: 122 (first correction in 42 BC, suggesting a rhythm of three instead of four years). Both interpretations lead to the same conclusion: there was no exact, written rule.

[60] Macrob. *Sat.* 1.12.34–5 etc.

Annalistic evidence also suggests a transfer of the regulation of intercalations to the *pontifices* in 191. The three extant mentions of intercalation in Livy begin in 190 BCE, and in each case occur in contexts suggesting that they derive from the *commentarii pontificum*.[61]

Examination of the politico-religious context provides further support for the hypothesis that Acilius' law was relevant in this regard, by showing that a definitive, new transfer of authority over intercalation to the *pontifices* in 191 would have been no anachronism, but a calculated and credible initiative.

Criticism of the transfer was founded on the opportunities for financial manipulation that it gave the *pontifices*, and our cited author, the Imperial-Period polyhistorian Solinus, associates this factor so closely with the measure that it is evident that he and his source had a single chronological context in mind.[62] The necessary basis for such manipulations of interest dates is an appropriately well-developed money economy, which cannot be looked for in Rome before the third century.[63]

In order to assess such economically based criticism of the intercalation process, we must first understand the situation at the time in respect of interest on debt, and especially the question as to whether interest was payable in the intercalary month. There is no well-founded evidence for such payments in (or for) the *mensis interkalaris*. Any transfer of Greek practice, which in fact made the intercalary month subject to interest,[64] is unlikely in view of the legal conception in Rome that the intercalary period was *momentum temporis*;[65] this distinction is the easier to understand as the Roman intercalary month was not a 'true' (lunar) month. If, then, we adopt the working hypothesis that it was interest-free, this produces an advantage for money-lenders, and especially for *publicani*. They received their taxation franchise at a fixed price on an annual basis, but in an intercalary year they were able to collect taxes outside Rome over a period of thirteen Roman months; further advantages may have arisen from the mismatch with local calendars that began a second year within that period.

The term 'money economy' also provides a hint of a terminological problem. In a small number of sources, the intercalary month is called *Mercedonius* or *Mercedinus*,[66] although in inscriptions and in dates mentioned in the literature it is always described as *mensis interkalaris*.[67] The latter circumstance is sufficient to enable us to reconstruct *Int.* or *Interc.* over the column for the thirteenth month in the Fasti Antiates maiores (the first publisher also added *Mer*);[68] but that does not explain the variant name. It is, however, striking that, of all the non-antiquarian

[61] See triumphs in 190, 170 and 167 (Liv. 37.59.2; 43.11.13; 45.44.3); RÜPKE 1993b.

[62] Solin. 1.43.

[63] CRAWFORD 1974; NICOLET 1980; DE MARTINO 1985: 62–5.

[64] LUCE 1967: 384 n. 2.

[65] *Dig.* 50.16.98.

[66] Plut. *Numa* 18.3; *Caes.* 59.3; Cinc. fr. 10 Funaioli = Lydus, *Mens.* 4.144.

[67] Fasti triumphales; Cic. *Quinct.* 79.

[68] *Editio princeps*: MANCINI 1921: 77. DEGRASSI (*Inscr. It.* 13.2.26): *Interk(alares)*.

authors, only the Greek Plutarch uses the expression. The answer must be that we have here a nickname for the intercalary month,[69] and that Plutarch, unlike the Latin antiquaries (and other 'native speakers'), did not see the joke. The term is not alone in being derived from *merces* ('wage', 'pension', 'payment', 'interest');[70] *mercedarius* and *mercedituus* ('breadwinner' and 'wage-slave') attest to analogously formed nicknames.[71] This less than laudatory coining may have been motivated by a connection with the manipulation of interest, made possible by the month regarded in law as representing a 'non-period'. We are reminded of the accusations against the *pontifices*: the nickname is unlikely to date from before the *lex Acilia*.

To conclude, I draw the reader's attention to two additional important indications linking Acilius' law with the *pontifices*, and in particular with Licinius Crassus. They concern firstly Licinius Crassus' widely famed financial genius, one of its manifestations – in public life – having been his appearance on the battlefield at Bruttium, and his acquisitive behaviour there: a trait that won him (and his descendants) the cognomen *Dives*.[72] He displayed the same skill, and the ability to exploit deficiencies in the calendar for monetary gain, in his handling of the *ver sacrum* in 194. This concerned the required repetition (*instauratio*) of the *ver sacrum* so praised during the Second Punic War, which was restricted to the sacrifice of all the newborn animals of one spring. The effect of this reprise of the 'sacred spring' was made less onerous by being precisely scheduled for the period between 1 March and 1 May, thus 'fortuitously' falling in December and January (by the Julian calendar).[73]

The second instance concerns the relations, implied by Solinus and his sources, and known to have existed, between the pontifical college and the *publicani*. In the year of his people's tribunate (212 BCE), Servilius Geminus had interceded for a *publicanus*, despite proven embezzlement; although it must be said that at this point he was not yet a *pontifex*.[74]

Such connections, at first sight apparently harmless, must be seen in the light of economic circumstances at the time. The stage reached by the money economy in Rome is illuminated by three accounts set in immediate chronological proximity to the *lex Acilia*. For 193, Livy mentions subterfuges for the avoidance of Roman laws against usury, involving allies, who were not affected by these laws, acting as 'front-men';[75] for 192 he reports usury trials resulting in substantial fines.[76] In 191 too, the year of the *lex Acilia*, a *lex Iunia de feneratione* – against usury – was adopted.[77]

69 Thus already MICHELS 1967: 18.
70 On Porph. *Hor. Sat.* 1.2.14 see NADJO 1989: 298–9.
71 See Sen. *Controv.* 10.4.5. 25; Paul. Fest. 111.18 L.
72 See Liv. 28.46.4–6.
73 See HEURGON 1957: 46–9.
74 Liv. 25.3.15. See BROUGHTON 1951: 271–2 n. 5; MALITZ 1987: 108.
75 On the *lex Sempronia* see DE MARTINO 1985: 172; cf. Plaut. *Curc.* 508–11.
76 Liv. 35.7.2; 35.41.9–10.
77 SCHLAG 1968: 158 n. 162.

If these hypotheses are valid, the *lex Licinia* and *lex Acilia*, two statutes passed five years apart, no longer merely represent evidence of a growing cult apparatus: they also attest to a policy pursued by particular individuals in respect of religious institutions, one of those individuals being Licinius Crassus, the first plebeian known to have been elected *pontifex maximus*. One of his henchmen was Acilius Glabrio. Like the calendar reform of Cn. Flavius and the *lex Hortensia*, the regulation of the calendar provided for in the *lex Acilia* is not of primarily chronographical interest: the way a society organizes its affairs in respect of time goes to the core of its political life.

7

Reinterpretation of the *fasti* in the Temple of the Muses

7.1 Marcus Fulvius Nobilior, *Triumphator*

If our reconstruction of the history of the *fasti* and the Roman calendar is valid so far, upon renewed consideration of the Late Republican calendar we encounter a substantial objection: where do the dedicatory notes come from, the data on *dies natales templorum*? Having no influence on the character of days, they have no function at all in the politico-legal instrument we have reconstructed up to now. What if we have been entirely mistaken in our characterization of the object of our endeavours? Perhaps, in the *fasti* of Cn. Flavius, we have all along been dealing with a regulatory instrument that also had functions in the religious field.

The fact that, even in the most lavish calendars, the selection of temple-dedication days appears arbitrary, and far from comprehensive, immediately makes this objection seem dubious. But the fact that they appear at all calls for an explanation. The solution should be sought in the form of a historical hypothesis: how did such information come to be included in the calendar? The answer is to be found in the context of that phase of momentous politico-religious developments already touched upon in connection with the *lex Acilia*, a generation after the end of the Second Punic War. Although, generally speaking, sources are more plentiful for this period, we have no direct witness for the event in question here. More than is true for preceding phases of our story, any date we may arrive at for this particular event thus takes on the character of a hypothesis. It is, however, vital that we look for that hypothetical date: it is 'extra touches' such as these that reveal the historicizing and sacralizing perspective of the *fasti*, so central to our wider historical understanding. In attempting this important task, we shall turn our attention to M. Fulvius Nobilior, the first man to whom we can ascribe the association of a written calendar with a sacral context: the publication of *fasti* in a temple.[1]

1 PIETILÄ-CASTRÉN 1987: 95–103; MARTINA 1981; ABERSON 1994: 199–216.

The Roman Calendar from Numa to Constantine: Time, History and the Fasti, First Edition. Jörg Rüpke.

Probably born soon after the First Punic War, Nobilior took to the political stage during the decade following the Second Punic War. *Aedilis* in 196 BCE, he became *praetor* in 193, and in the two following years distinguished himself militarily in Spain. In 189 he was consul with one Manlius Vulso, and assumed military command on the Aetolian front.[2] Tradition records as the most important event of this campaign the siege and capture of the city of Ambracia, north of the Gulf of Actium, and it was in fact the booty from this city in particular that the victor displayed in his triumphal procession after his return in 187.[3] Both Livy's description of the triumph and later accounts demonstrate that Fulvius, with C. Claudius Marcellus, the victor over Syracuse, stood at the beginning of that series of Roman commanders who, in giving expression to the Roman desire to emulate Greek and Hellenistic world culture, augmented their wars by the added dimension of the organized plundering of works of art.[4] Dubious as such testimonies to Fulvius' artistic sense may seem, it is confirmed by his subsequent actions. Fulvius not only brought the treasures home: he put a roof over their head. He transformed a temple of Hercules into a 'museum' in the real sense of the term: a *Museion*, a sanctuary of the Muses, and the first in Rome.[5]

The details of the structure are disputed in two respects. The first problem is the date. There is no evidence that Fulvius vowed the temple during his campaign, even before Ambracia: any such vow would in all events have been realized shortly after his triumph.[6] In fact, the only evidence to make any explicit suggestion of a date speaks of the structure's being funded by a censorship: but Fulvius was not a censor until a decade later, in 179 BCE, a campaign five years previously having failed.[7] The funding detail comes from a unique source, and its chronological distance from the event in question – Eumenius brings up the matter in a panegyric delivered in 297 CE – is not the only factor that permits us to reject it.

The great construction programme, undertaken by the censorate served in common with the *pontifex maximus* and *princeps senatus* M. Aemilius Lepidus, and also intended as an act of appeasement,[8] may very well have employed agreed Senate funds to serve highly private intentions.[9] Finally, it remains conceivable that private means were used for a project begun during the censorate.[10] Striking,

[2] Liv. 37.50; 38.1–11.

[3] Liv. 39.5.14ff.

[4] Pape 1975; Östenberg 2009.

[5] Stressed by Feeney 2007: 144.

[6] Aberson 1994: 205–6. The *manubiae* mentioned in Cic. *Arch.* 27 must be no more than representations of the Muses themselves (thus already Shatzman 1972: 182).

[7] Eumenius, *Paneg. Lat.* 9.7.3: *Aedem Herculis Musarum in circo Flaminio Fulvius ille Nobilior ex pecunia censoria fecit, non id modo secutus quod ipse litteris et summi poetae amicitia duceretur, sed quod in Graecia cum esset imperator acceperat Heraclen Musageten esse, id est comitem ducemque Musarum, idemque primus novem signa, hoc est omnium, Camenarum ex Ambraciensi oppido translata sub tutela fortissimi numinis consecravit, ut res est, quia mutuis opibus et praemiis iuvari ornarique deberent: Musarum quies defensione Herculis et virtus Herculis voce Musarum.*

[8] Münzer 1910: 267, 1920: 200–1. See also Liv. 40.46.

[9] Martina 1981: 53; followed by Skutsch 1985: 145 n. 3.

[10] Pietilä-Castrén 1987: 101; Ferrary 1988: 566.

however, is the fact that Livy does not mention the temple in his comprehensive description of the censor's construction activities:[11] this reservation also applies if Livy was writing in the light of the lavish restoration of the Temple of the Muses by L. Marcius Philippus,[12] after the latter's Spanish triumph of 33 BCE, and when Ovid, in his commentary on the *fasti*, had made Marcius alone responsible for having associated the Muses with Hercules.[13] On the other hand, Livy speaks of the construction of a *porticus* at a temple of Hercules in the context of Fulvius' building programme as censor.[14]

A further factor suggesting a late date under the censorate for the building project is Fulvius' relationship with M. Porcius Cato. The latter attacked him in speeches during his year in office, and criticized him for his use of the booty, the sources citing the speech *Uti praeda in publicum referatur*.[15] The positioning of the plundered statues at a public location, the Circus Flaminius, that recalled the triumphal procession, and established connections with the fulfilment of his colleague's equally long-standing wartime vow,[16] may have represented a reaction to Cato's attacks.[17]

The problem may be resolved if we pursue the added complexity of the precise character of the dedication. While many Imperial-Period sources speak of an *aedes Herculis Musarum*, some from the Republican period, or when discussing Republican themes, mention an *aedes Musarum* or *Camenarum*. Servius adds the further complication that Fulvius installed into an existing temple of Hercules an *aedicula* of the Muses or Camenae dating back to Numa.[18] This last possibility may be rejected immediately: the description of the structure as *aedes Herculis et Musarum* is found only in this one source, and it appears improbable from the standpoint of sacral law that the dilapidated *aedicula* should have been moved twice; nowhere are we told of an older temple of Hercules on the Circus Flaminius near the temple of Hercules Magnus Custos.[19] The first-mentioned point has given rise to the surmise that the introduction of Hercules Musagetes into the cult of the Muses could represent a Late Republican or Augustan innovation.[20] But this founders on the character of the various sources. First: the term *Hercules Musarum* is used by those very authors who offer the most precise information on the circumstances of

[11] ROSEN 1985: 85 n. 94.

[12] Suet. *Aug.* 29.5; see RICHARDSON 1977: 359–61.

[13] Liv. 40.51; Ov. *Fast.* 6.799–802; GROS 1976: 22. Cf. MARTINA 1981: 54.

[14] Liv. 40.51.6.

[15] Fragments: Cato, *Orat.* 72 Sblendorio Cugusi = Prisc. *Gramm.* 2.368.1–3 Keil: *Miror audere atque religione non tenerei, statuas deorum, exempla earum facierum, signa domi pro supellectile statuere.* See also Cato, *Orat.* 103–6 Sblendorio Cugusi; cf. SBLENDORIO CUGUSI 1982: 293–9.

[16] See Liv. 40.52.

[17] MARTINA 1981: 58–65.

[18] On the number of the Muses, Serv. *auct. Aen.* 1.8 explains: *His Numa aediculam aeneam brevem fecerat, quam postea de caelo tactam et in aede Honoris et Virtutis conlocatam Fulvius Nobilior in aedem Herculis transtulit, unde aedes Herculis et Musarum appellatur.*

[19] On this critique MARTINA 1981: 51; see also TAMM 1961: 165–6.

[20] Thus RICHARDSON 1977.

Fulvius' foundation: Eumenius, who gives us the source of funding, and Macrobius, through whom we have the *fasti* dedication. Second: in the cases that omit to mention Hercules, there are understandable motives for their concentration on the Muses, as the focus of these passages is on the poets assembled in the temple.[21]

In the light of the problems and ambiguities involved both in dating the construction of the temple and in the name, all the evidence appears to point to the following hypothesis:[22] Fulvius neither vowed nor constructed a temple. Any intentions he may have had in that direction were countered by senatorial and pontifical criticism of his plundering of Ambracia.[23] It was only after becoming reconciled with Aemilius, his former antagonist and now fellow censor, that he undertook the extension of an existing temple of Hercules, probably that of Hercules Custos. He provided it with a columned hall, and then installed the Ambracian Muses there. Hercules became Hercules Musarum. *Terminus post quem* is the beginning of Fulvius' censorate in 179; completion and dedication would have occurred in the years following.[24]

To understand the important role of Hercules, we have only to consider the inherent symbolism of the temple building.[25] The least of the motifs involved is the close association of the Roman Hercules with victory and triumph.[26] Much more important is the interpretation, or perhaps authentic tradition, handed down by Eumenius as the message symbolized in the temple building, in the passage already cited: only Hercules' protection affords the repose necessary for the Muses' creative unfolding; the voice of the Muses proclaims Hercules' courage. The expression of this symbolic relationship corresponds – in Fulvius' interpretation – to the real relationship between himself, the commander, and Ennius, the poet who accompanied him on his campaign, and whose works – in particular the (dramatic?) poem *Ambracia* – proclaim Fulvius' glory.[27]

[21] See TAMM 1961: 157–9, 1969: 325.

[22] ABERSON 1994: 211–16 following CASTAGNOLI.

[23] See Liv. 38.44; cf. GRUEN 1992: 110.

[24] FEENEY's arguments (2007: 143, following GRATWICK) are valid for the importance of 184 BC in Ennius' *Annales*, not for a foundation date of the temple.

[25] MARTINA 1981: 63–8; GRUEN 1984: 259 n. 52.

[26] See RÜPKE 1990a: 194. 210 with additional literature.

[27] See SCHANZ AND HOSIUS 1927: 88, with sources, in particular Cic. *Tusc.* 1.3: *oratio Catonis, in qua obiecit ut probrum M. Nobiliori, quod is in provinciam poetas duxisset; duxerat autem consul ille in Aetoliam, ut scimus, Ennium.* See also *Vir. ill.* 52.3: ... *quam victoriam per se magnificam Q. Ennius amicus eius insigni laude celebravit.* As already determined by the Republican grammarian L. Aelius Stilo (see Gell. 12.4, where the Ennius quotation may also be found), Ennius introduced his relationship with Fulvius into his literary output, in his *Annales*, in the shape of a military commander in the Second Punic War, Cn. Servilius, accompanied by a poet (thus SKUTSCH 1985: 448; support for Stilo in LEO 1913: 161 and NORDEN 1915: 133): Enn. *Ann.* 268–86 Skutsch = 234–51 V. The association of Hercules with the Muses already existed in Greece, and Fulvius may have derived the juxtaposition from there: see BÖMER 1958: 390 and GRAF 1985: 98 for supporting evidence; against both these authors, however, it must be pointed out that, in Greece, Hercules was dominant, and the association with the Muses was frequently placed in an athletic context, while in Rome the Muses took on the dominant role, and poetry moved to the fore.

Early on, there were certainly oral traditions, stories and songs celebrating the heroic deeds of ancestors.[28] From the beginning of the third century onwards, we have evidence of honorific funerary inscriptions, increasing in scale with the *elogiae* to the Scipiones;[29] our earliest funerary oration (*laudatio funebris*), in fragmentary form, and thus presumably a speech recorded in writing, dates from 221; it concerns the twice consul and – absent from the textual fragment – *pontifex maximus* L. Caecilius Metellus.[30] Other genres, more relevant to our present purpose, had their origins during this period: the first speech to come down to us in written form was given by Ap. Claudius Caecus in 280;[31] and, in the middle of the century, the *pontifex maximus* Ti. Coruncanius began to record pontifical *commentarii*.[32] Both had a mainly legitimatizing function.

In this same line of development, when Fulvius adopted a common Hellenistic practice by taking Ennius with him to Aetolia,[33] his action represented an important stage in Rome's cultural history. In spite of critical voices,[34] the profession of literature, having had a place in Rome since Livius Andronicus, was at this time definitively adopted into the repertoire of aristocratic rivalry;[35] following a path prepared by Naevius and the comedies of Plautus, it finally acquired a public, political presence on the sacral stage.[36]

That this was an eminently political process is demonstrated by the Cato quotation passed down by Cicero: in the same context in which he criticized Fulvius for taking Ennius with him on his campaign, Cato asserted that even the old heroic songs – a genre that had, in the meantime, become obsolete[37] – had been held in no great esteem. This indicates a context of actual controversy, with which we are made familiar a good one hundred years later in Cicero's letter to Lucceius. If the historical reporter's assessment is to meet with general recognition, it is better by far that he should not be the same as the person being reported upon:[38] the more independent the judge, the more credible the verdict. Cato's opinion should be viewed against such a background. He does not really take up an opposing position, but – as is demonstrated by his own reflections on the disclosure of names in the *Origines* – takes the exploitation of literature, as practised by Fulvius, as a given common basis for his argument. This was the basis for his historical

[28] MURRAY 1985; BREMMER 1990; ZORZETTI 1990; PERUZZI 1993.

[29] COARELLI 1972.

[30] Plin. *HN* 7.139–41; KIERDORF 1980.

[31] Cic. *Brut.* 61.

[32] See RÜPKE 2008: 24–38.

[33] See e.g. BURCK 1992: 22 on the two historians Sosylus of Sparta and Silenus of Caleacte who accompanied Hannibal (Nep. *Hann.* 13). On the importance of the military theme in Hellenistic court poetry, and its differentiated development after Alexander, see WEBER 1993: 302–19.

[34] See ROMANO 1990: 98–9, 106. Cf. Plautus' *Miles Gloriosus* (25ff.).

[35] See RÜPKE 1995b.

[36] See RÜPKE 2001.

[37] Cato's *Orat.* 104 Sblendorio Cugusi is made to use the perfect tense.

[38] Cic. *Fam.* 5.12, §§ 8–10; RÜPKE 1992a: 207.

judgement that the literary product had been discredited to the point of having become a system-immanent tool.

These literary and sociological developments did not occur in isolation. In the context of her rapid expansion as an international power since the middle of the fourth century, Rome was undergoing cultural adaptation and active integration into the Hellenistic Mediterranean world, a process that was further accelerated by her active incorporation of and dominance over that same Mediterranean world, especially after the First Punic War.[39] For the Romans themselves, that is to say the leading elite, the experience was one of deep-reaching modernization, explicitly inspired by Greek models.[40]

The same process is illustrated in many aspects of the circle involved in Fulvius' temple foundation. One of its members was the poet Q. Ennius, born in 239. His oeuvre gave further impetus to the reception of Hellenistic genres in Rome, and at the same time strengthened Greek traits in those genres that were already naturalized. In the *Euhemerus* he wrote the first philosophical treatise in the Latin language. As a 'philologist' of wide-ranging ambition (*dicti studiosus*),[41] he engaged in literary exegesis and introduced abbreviations and double consonants.[42]

The position is more palpable still in respect of Fulvius himself. Probably in preparation for games he had vowed during the Aetolian campaign, he brought many Greek artists to Rome, and presented athletic competitions to Roman audiences for the first time;[43] the exotic element was reinforced with an animal hunt (*venatio*) involving lions and panthers. Livy, who provides this information, maintains that, in size and variety, the spectacles presented rivalled those of his own time.[44]

The temple-extension project too has a thoroughly Greek feel. The very concept of Hercules Musagetes came from Greece. Ambracia was regarded as the city of Herakles.[45] The architecture is harder to assess. As a basis, we have only the Severan *Forma urbis*, and, while this provides sufficient fragments for a reconstruction of the Augustan building, it offers no reliable clues that would date the unusual *cella* structures back to the Republican Period.[46] On the other hand, the furniture of the *aedes* was undeniably of Greek origin. The statues of the nine Muses (although these may just possibly have been paintings) came from the Ambracian booty.[47] A series of coins of Q. Pomponius Musa, minted in Rome in 66 BCE, shows, besides

[39] GRUEN 1984: 252; detailed account 250–72.

[40] VEYNE 1979: 27; MACMULLEN 1991.

[41] Enn. *Ann.* 209 Skutsch = 216 V; SUERBAUM 1968: 270–80.

[42] SCHANZ, HOSIUS 1927: 91, 98.

[43] FERRARY 1988: 519–20.

[44] Liv. 39.22.1–2: *†Decem apparatos deinde† ludos M. Fulvius, quos voverat Aetolico bello, fecit. multi artifices ex Graecia venerunt honoris eius causa. athletarum quoque certamen tum primo Romanis spectaculo fuit, et venatio data leonum et pantherarum, et prope huius saeculi copia et varietate ludicrum celebratum est.* GRUEN 1984: 260, 1992: 195–6; MACMULLEN 1991: 421–2.

[45] BURKERT 1961: 242–3 n. 3.

[46] TAMM 1961: 162–3; FERRARY 1988: 568.

[47] Plin. *HN* 35.66 on Zeuxis: *Fecit et figlina opera, quae sola in Ambracia relicta sunt, cum inde Musas Fulvius Nobilior Romam transferret.*

the nine Muses, Hercules playing on a lyre;[48] this could indicate a corresponding statue from the same source.

Finally, Marcus Fulvius Nobilior may have adopted a high-profile stance in his adaptation of Hellenistic culture,[49] but it must once more be emphasized that he was not an isolated figure: in 180, the same year in which he was elected censor for the following year, another Fulvian, Q. Fulvius Q. f. M. n. Flaccus, attained the pontificate, and was elected consul, also for 179.[50] Even more indicative of the esteem Fulvius Nobilior enjoyed among the great and good of the ruling elite, however, is the fact that, at the same time, his son Quintus was co-opted as a member of the *epulones*, founded some fifteen years beforehand: and this before assuming the *toga virilis*; thus as a minor, so to speak.[51] In the general scheme of things, Fulvius did not even differ from his adversary[52] Cato: in the appropriation of Greek artefacts and other elements of the Greek cultural heritage, which had always occurred, but was now pursued with heightened intensity, the Roman nobility manifested their self-assumed prerogative to exploit a role that was increasingly that of a 'world power'.[53] It must be said that any idea of sparing the Greeks, either militarily or in terms of power politics, had no place on the agenda of these philhellenists.[54]

Discussion of the interior of the temple has up to now omitted the detail that has most bearing on the history of the Roman calendar. We know of it through the pen of only one, late, but at the same time very valuable source: Macrobius. In discussing the names of the months, he asserts that the interpretation of *Maius* was much disputed. He opens the long list of different positions with a quotation, which he introduces thus: 'Fulvius Nobilior, in the calendar *(in fastis)* which he deposited in the temple of Hercules "Leader of the Muses", says ...'.[55] What is Macrobius talking about here? What kind of *fasti* are these? Unfortunately, the verb used, *ponere*, fails to provide an answer: it could apply to any conceivable material, any conceivable object. We are left with the term *fasti*.

Discussions have centred around two possibilities: Fulvius dedicated a painted calendar, or a book. Assuming a two-dimensional calendar, other media may be excluded. Bronze calendars are not attested in Rome, and we would expect a large inscription to be on the external wall. A calendar as a material object would seem an unlikely subject for a dedication. Calendars had already been in use a long time, so Fulvius' specimen was not an original, like the standard metre in Paris. To draw a parallel chronologically closer to our problem, it did not merit preservation in

48 *RRC* 410/1 and 410/2–10.

49 PAPE 1975: 59; PIETILÄ-CASTRÉN 1987: 163; RICHARDSON 1991: 392–402.

50 Liv. 40.42.11; RÜPKE 2008, no. 1763.

51 *Ibid*., no. 1771.

52 See GRUEN 1984: 260–6; cf. MACMULLEN 1991: 429–38.

53 VEYNE 1979: 9.15.

54 GRUEN 1984: 267–70.

55 Macrob. *Sat.* 1.12.16: *Nam Fulvius Nobilior in fastis, quos in aede Herculis Musarum posuit, Romulum dicit* ... Tr. P. V. Davies 1968.

the same manner as treaties or statutes. It thus appears improbable that this was a costly specimen.

There remains the possibility of a book.[56] The mere fact that statements undeniably of the nature of a commentary are cited from Fulvius' *fasti* is not sufficient to compel the assumption that the text itself took the form of a treatise: Verrius Flaccus' Fasti Praenestini demonstrate that a commentary can be accommodated even within the material bounds of the *fasti*. Regrettably, our knowledge of *fasti* in book form is sparse. Any attempt at reconstruction can no longer credibly rely on the *laterculus* of Polemius Silvius and its *fasti*, although the structure of the original manuscript of 449 CE may conceivably have been similar to that of the 'Chronograph of 354'. For all its abundance of material, it must be said that the Chronograph embodies nothing that may strictly be termed a commentary: it offers only the 'documents' themselves. Ovid's calendar poem does not belong to the *fasti* genre: its title *Libri fastorum* shows it to be a commentary, and should be understood in the sense *De fastis*, not *Fasti*.[57] Taken together, these considerations lead one to conclude that the idea of Fulvius' depositing a mere calendar roll in his temple may be dismissed as laughable, and that any other case of an object in book form could not be described simply as *fasti*; *librum, quem de fastis scripserat* would be the least we might expect of Macrobius by way of a more specific definition. The references included in his brief section on the calendar show that Macrobius' sources, and he himself in particular, give very precise indications of the genres and titles of the texts they (indirectly) use; there are no grounds for assuming that they were negligent in the particular case of the work *in aede Herculis Musarum*.[58]

These considerations also conflict with the assumption that there was not only a painted calendar, but also an accompanying commentary (or a commentary published only secondarily as a painting).[59] Examination of the few statements ascribed to Fulvius concerning the calendar[60] reveals that all quotes go back to M. Iunius Gracchanus, who wrote after the middle of the second century BCE. There is therefore nothing to indicate the circulation of a text by Fulvius in book form. Iunius himself was writing *de potestatibus*,[61] thus was not continuing a genre *de fastis* initiated by Fulvius.

[56] PAIS 1909: 208–9; FRAZER 1929: 4.346; BOYANCÉ 1955: 174; BURKERT 1961: 241.

[57] RÜPKE 1994; RÜPKE 2009b: 131–7.

[58] *Cingius in eo libro quem de fastis reliquit* (1.12.12); *sicut Nisus in commentariis fastorum* (1.12.30); *Glaucippus qui de sacris Atheniensium scripsit* (1.13.14); *Antias libro secundo* (1.13.20); *Tuditanus refert libro tertio Magistratuum* (1.13.21); *ut Cato in Originibus* (1.14.5); *Plautus in Curculione* (1.16.14); *sicut Varro in augurum libris* (1.16.19); *Gellius Annalium libro quinto decimo et Cassius Hemina Historiarum libro secundo* (1.16.21); *Fabius Maximus Servilianus pontifex in libro duodecimo* (1.16.25); *Q. Claudius Annalium quinto* (1.16.26); *Titus de feriis scribens* (1.16.28); *Trebatius in libro primo Religionum (ibid.)*; *Iulius Caesar sexto decimo Auspiciorum libro* (1.16.29); *Cornelius etiam Labeo primo Fastorum libro (ibid.)*; *apud Granium Licinianum libro secundo* (1.16.30).

[59] MICHELS 1967: 125 n. 18.

[60] RÜPKE 2006.

[61] MERCKLIN 1840: 43–4.

The above reflections only underpin what would in any event be the simpler solution in respect of a dedication made by a portico-builder who did not go on to be well known as a literary figure: Fulvius installed *fasti* in the *aedes Herculis Musarum* in the form of a wallpainting.[62] For aesthetic motives too – although, admittedly, such as would quickly become anachronistic – a painting of this nature fits better in the interior of a meeting place for poets, furnished with many statues and (or?) Greek paintings, especially as the further Alexandrine idea of combining a library with the *Museion* was not[63] adopted.

How, then, to explain the propositions ascribed to Fulvius' *fasti*? Even though Fulvius is cited for a few etymologies of month-names, the common purport of the propositions cited does not lie in the explanations they give for some elements of the calendar, but in their elucidation of its history: the etymologies of March and April (Mars and Venus), and those that follow them, are interesting because they describe Romulus as the calendar's creator, and Numa as the man who completed it; even Acilius' calendar law features among the information cited. We know these propositions only as restated by Iunius. In condensed form, they could well have found space enough in a lengthy caption, functioning simultaneously as a dedicatory inscription, and thus also containing Fulvius' name.

Against this background, the way the *fasti* were received by their audience acquires a quite particular perspective: they become a document that might provide a reading of history. But this historical information does not lie in the feast-names, and resides only to a slight extent (clarified by Fulvius in the caption) in the names of the months. The approach already has a distanced, antiquarian character. It finds comprehensive expression in the Varronian nostalgia of the Late Republic; it typifies the Augustan Period, and may be discerned in the commentaries of Ovid and Verrius Flaccus. But the audience targeted by Fulvius' historicization was of a different kind. In that half-century that followed the end of the war against Hannibal, in the eastern Mediterranean especially, the Romans were busy gathering in the harvest they had been preparing for during one and a half centuries of armament and power politics; in Rome, temples financed from booty were shooting up like mushrooms; Romans then saw history differently, in much more tangible terms, in those very temples. And, in Rome, temples had to do with calendars; more precisely: with the temple-foundation days in the *fasti*.

7.2 Temple Dedications in the *fasti*

The answer to the question of the origin of the dedication notices in the *fasti* has a secure *terminus post quem* in the creation of the *fasti* by Cn. Flavius. There appears

62 Thus also DEGRASSI 1963: xx.

63 RAWSON 1991: 442.

to be no convincing reason why a calendar whose purpose was to document the political and juridical year, and which included religious dates only where they affected (or, unexpectedly, did not affect) that purpose, would record *dies natales templorum*. A safe *terminus ante quem* is provided by the Fasti Antiates maiores from the sixties of the first century BCE, where such entries are certainly present. Like all later *fasti*, they only provide the dates for temples in the city of Rome. It follows that the *fasti* from Antium were based directly or indirectly on a model in or from the city of Rome. This brings the *terminus ante quem* further forward, and suggests that the nature of that model is to be sought in a relationship to the wall-calendar in the Museion. Only Antium, however, can provide the key.

Analysis of individual calendar specimens[64] reveals selectivity in respect of the dedication notices included. That selectivity, especially when very rigidly applied, often betrays the principles underlying it; but in many cases it appears quite arbitrary. The same does not apply to the Fasti Antiates, where the available 'selection' of temples is very extensive: the condition of the *fasti* so astutely assembled by the first editor is so bad, despite the impressive work of reconstruction, that nearly all foundation days otherwise attested as chronologically possible for inclusion in the Republican *fasti* would fit into the missing portions.[65] This also applies in the case of the Aventine temple of Consus, probably built by L. Papirius Cursor after his triumph in 272 BCE. Its foundation date is given in the Fasti Vallenses as 21 August, a day that is well preserved in the Antiates maiores, but without this reference;[66] the Fasti Ostienses and Amiterni locate the same occasion on 12 December, which is probably correct, although this date is not extant in the Antiates. The reason for the incorrect date is probably the Consualia on 21 August:[67] no necessary connection in fact exists between this ancient festival and the temple.

Augustan restorations are not only a fundamental source of uncertainty; they also provide an opportunity to elucidate the entire question. In the Res Gestae, Augustus speaks of eighty-two restored temples, and numerous other restorations by other members of the ruling elite are attested. In principle, we must assume a redefinition of the *dies natalis* for each of these restorations.[68] That this hypothesis should nevertheless be treated with care is shown by the dedication of the temple of Jupiter Victor on the Quirinal, shortly after it had been vowed in 295 BCE, attested for 13 April by Ovid and by the Fasti Antiates maiores.[69] The temple of Jupiter Invictus mentioned by Ovid for 13 June, and missing from the Republican calendar, is not an identical doublet, arising from a rededication, but very probably dates from the second or final third of the second century BCE.[70] Taking all the

64 RÜPKE 1995a: 39–164.

65 Not indicated in DEGRASSI's edition.

66 *Inscr. It.* 13.2.149.

67 *Inscr. It.* 13.2.106. 199; ZIOLKOWSKI 1992: 24.

68 August. *Gest.* 20.

69 Ov. *Fast.* 4.621–2 (13 April).

evidence into account, it is thus possible to surmise that the record in the Fasti Antiates is nearly complete.

Qualification as 'complete' can, of course, relate to only one particular point in time: temple-foundation days can become obsolete due to the physical collapse of a temple; newly built temples can be added after the calendar's completion, or remain unheeded. The former aspect must remain a closed book, owing to the limited body of source material available to us; the latter consideration, however, leads from the Fasti Antiates, which show no trace of late additions, to the question of our original Roman model.

As has emerged already in our discussion of the calendar from Antium, the only source to provide secure identifications is the consul list. Analysis of data concerning temples is severely impeded by the lack of annalistic writings after the second third of the second century: Livy's record ceases in 167 BCE. What is certain is that Antium's list of recorded temples extends until 173 BCE. In this year on 13 August, the temple of Fortuna Equestris was dedicated by Q. Fulvius Flaccus, one of the two censors for that year:[71] the same 13 August that marks the foundation day of Fulvius' Museion. The further names listed between these two temples in the Fasti Antiates may, therefore, likewise relate to contemporary structures, and cannot be used for establishing earlier dates.

No secure later dates can be identified. The Lutatius Catulus[72] who built the temple of Juturna (11 January) may well have been the consul of 102, but the building should more probably be ascribed to the proconsul and *triumphator* of 241.[73] Identification of the *[—] Invict(o)* on 15 May with the Mars Invictus attested in the Imperial-Period Fasti Esquilini for 14 May is highly problematic: there is no conceivable reason why the dedication should have been wrongly assigned to the Ides or the day before the Ides. Identification of this temple with the temple of Mars built in 138 BCE by the consul for that year, D. Iunius Brutus Callaicus, is, again, pure supposition.[74] The temple of the Lares noted on 27 June in the Fasti Antiates maiores is first attested in 106 BCE on the occasion of a prodigy; but it may have been constructed centuries earlier.[75]

Although integration of the temple of Felicitas on the Capitol, to cite the more complete information given in the Fasti Antiates ministrorum, with its putative dedication on 1 July, is probably correct, it is inadmissible to identify this temple with the Sullan temple of Fausta Felicitas, whose *dies natalis* in the Tiberian period lies on 9 October: why would the Tiberian Fasti Antiates ministrorum record the earlier foundation day, and under another name?[76] It remains true, however, that other identifications that could also lead to a more recent date are lacking.

70 Ov. *Fast.* 6.649f.; *Inscr. It.* 13.2.103 (13 June); ZIOLKOWSKI 1992: 80–1.

71 Liv. 42.10.1–5.

72 Serv. *auct. Aen.* 12.139.

73 DEGRASSI 1963: 395; COARELLI 1975: 252; ZIOLKOWSKI 1992: 94–7.

74 Against DEGRASSI 1963: 459, 457.

75 Obseq. 41; DEGRASSI 1963: 474; ZIOLKOWSKI 1992: 98 (third century BCE).

76 Against DEGRASSI 1963: 475. *Inscr. It.* 13.2.208.

Uncertainty also surrounds the dedication notice for a temple of Concordia repeated in the Fasti Pinciani on 22 July.[77] There probably never was a temple going back to Camillus.[78] The relevant foundation day in Ovid (16 January) should be assigned to Tiberius' rededication of the temple dedicated by L. Opimius, consul in 121 BCE; the Republican date thus remains open. But the reference may also be to the *aedes Concordiae* founded by Cn. Flavius, back in 304.[79]

The identification of the sanctuary of Jupiter Stator, whose *dies natalis* the Fasti Antiates maiores give as 5 September, also appears problematic. DEGRASSI relates it to a building of Q. Caecilius Metellus in 146 BCE.[80] But another possibility is the structure associated with M. Atilius Regulus, consul for 294, supposed to have replaced a simple *fanum* dedicated to the god by Romulus.[81] Ovid gives the Imperial-Period foundation day as 27 June. Although there is no explicit evidence for a new dedication, the Antiates maiores do not mention Jupiter on this day.[82] If Ovid's information is true, then the *dies natalis* of 13 January given by the Fasti Filocali perhaps represents the foundation day of Marcellus' sanctuary, or a renovation.[83]

Our final candidate is the temple of Fides, dedicated on 1 October. Its cult is traced back to Numa; the *aedes* dating from the First Punic War was renovated or enlarged at the end of the second century BCE by M. Aemilius Scaurus, victor over the Cimbri.[84] In view of the lavish ritual expended once a year in the Fides cult, and therefore capable of being associated with the *dies natalis* of the temple,[85] the dedication day on the October Kalends, which is also found in later calendars, may have survived the not particularly thoroughgoing renovation:[86] it was not so easy for a Republican patron to deflect the habits of priests, in this case presumably the three great *flamines*, accustomed to driving to the Fides in a carriage in order to carry out their rite with covered hands.[87]

To summarize the situation once again, all the extant dedication notices of the Fasti Antiates maiores can be related to temple constructions prior to 173 BCE. This finding is subject to the caveat that, from the debris of the wall-calendar from Antium, a substantial number of entries have not survived. The list of those remaining is nevertheless of considerable length: at some seventy-five items, it undoubtedly represents the greater proportion of all conceivable *dies natales*. The evidence may therefore be regarded as significant.

77 *Inscr. It.* 13.2.47.

78 ZIOLKOWSKI 1992: 22–3.

79 Liv. 9.46.6; DEGRASSI 1963: 486. 16 January: Ov. *Fast.* 1.637–44; WISSOWA 1912: 328.

80 DEGRASSI 1963: 508.

81 Liv. 10.37.15; see also Cic. *Cat.* 1.33.

82 Ov. *Fast.* 6.793–4; ZIOLKOWSKI 1992: 88.

83 *Inscr. It.* 13.2.239.

84 See Cic. *Nat. D.* 2.61.

85 LATTE 1960: 237.

86 See COARELLI 1975: 45–6.

87 Liv. 1.21.4.

A line may now be drawn connecting this particular finding with Fulvius' *fasti*. When, in our discussion of temple construction so far, we have mentioned the year of the 179 BCE census, it has been only as a *terminus post quem*, before which the construction programme initiated by the censors could not have been concluded. The most definite chronological indication, the position of the dedication notice *Camenis* – why it could not be called *Herculi Camenarum* is explained by its construction history – *after* a temple already consecrated in 173 suggests a date in this year or after. Fulvius probably lived beyond 173; Ennius certainly did so.

Under these circumstances, however, it is no longer necessary to assume a specific link connecting Fulvius' *fasti* with the Fasti Antiates. This has implications: if a Latin wall decoration, in representing such a common instrument as the *fasti*, is modelled on another specimen in the city of Rome, one hundred and ten years older, we have the right to suspect that the 'fashion-setting' influence of that original specimen was not confined to Antium, but also operated in Rome itself. In any event, all later calendars follow the Fasti Antiates in their most important details.[88]

Temple foundations rank highly in the list of those historical events that were the most precisely retained in Rome's collective memory,[89] the characteristic forms of which will be a theme in what follows. Temples were monuments to the historical events that led to their foundation, and this applies especially to victories won in dangerous battles, often only thanks to divine aid.[90] Our evidence for this does not rely solely on the ability of the annalistic tradition to perpetuate the historical events in question. The converse premise produces the same outcome: significant defeats, for which no-one thanks the gods with temples, were likewise remembered.[91] That in the one instance the day of dedication is recorded, in the other the actual day of the battle, relates to differing sacral responses to the alternative outcomes, and will merely be noted here as a reality. That Fulvius had mastered the language of temple construction, and the art of choosing a dedication date, is shown in the case of the 'temple of the Camenae': the date in question is the Ides of August, when there is no festival to divert attention, but five other *dies natales templorum* form a sufficiently illustrious setting.[92] Moreover, the strategy was freshly tested. M. Aemilius Lepidus, the *pontifex maximus*, and Fulvius'

[88] That the model for the Fasti Antiates maiores dated back to Ennius at the earliest is suggested by the double consonants attestable only in the 'small entries' on 15 March (*Ann(ae) Perennae*) and 18 July (*Alliens(is) die(s)*), and supposed to have been introduced by Ennius (Festus, *Gloss. Lat.* 374.7–11 L); on his possible motivation (prosody) see SCHANZ AND HOSIUS 1927: 91.

[89] ROHDE 1932; RÜPKE 1990a: 234; RÜPKE 2006a.

[90] KEYSSNER 1935: 1800–1.

[91] BELLEN 1985: 23; GRAFTON AND SWERDLOW 1988; RÜPKE 1990a: 26–7.

[92] *Inscr. It.* 13.2.16: *[Eidus.] N[P.] Dianae Vortu(mno), Fort(unae) Equ(estri), Herc(uli) Vic(tori), [Cas]t(ori) Poll(uci), Came(nis).* The association between the Camenae and Fulvius' temple contradicts DEGRASSI 1963: 496. By Ovid's testimony, the foundation festival celebrated on 30 June in the Augustan period relates unambiguously to Philippus' restoration. The *natalis Musarum* attested in the Fasti Polemii Silvii for 13 June (*Inscr. It.* 13.2.269) cannot possibly be seen as a recollection of a pre-Augustan date.

colleague as censor, dedicated a temple of Diana in 179 on 23 December, and a temple of Juno Regina on the day of the Larentalia, which was already distinguished by the *dies natalis* of a temple of the Tempestates vowed by L. Cornelius Scipio during the First Punic War.[93] Just on the previous day,[94] however, Aemilius had created a further *dies natalis*, with the temple of the Lares Permarini.

These factors point to the possibility that Fulvius could have been the one who first recorded dedication notices in the calendar, creating for the temple foundations a showplace that did not depend on viewing the buildings themselves. His historicizing influence on the *fasti* is attested beyond doubt by the fragments, and he would have had a model in the person of his colleague Lepidus, who, in the case of his temple of the Lares, had not only affixed a substantial dedicatory inscription on the building itself, but had a copy hung on the Capitoline temple of Jupiter.[95]

Such an action would have been entirely in line with the notion of the calendar as a form of central showcase for the temple foundations of individual *nobiles*. But it must be remembered that Fulvius did not display his list of honour on the Capitol (as Lepidus did the copy of his own inscription) or the Forum (as Augustus did later with his triumphal *fasti*): if Fulvius' intention was indeed public display or social prestige, the method he chose was an indirect one; it must also be said that the picture of the *fasti* presented in the presumed caption was of a work whose historical coverage was quite extensive. The desire not only to make history oneself, but to represent history through the medium of one's own actions and those of one's ancestors, may be seen as a Greek idea, but the conception of history in this concrete form was Roman: the accomplishments of the Roman *nobiles*, made visible here, had made Rome what it was. Or, in the words of Ennius: *Moribus antiquis res stat Romana virisque.*[96]

The name Ennius reminds us not only of the audience for Fulvius' *fasti*, but of the refracting lens of historiography, through which historical events could be restored their original prestige: in Rome at this time, and in the Latin language, this literary form still appeared possible only in poetry. To install a wall-painting depicting a symbolic digest of Rome's history through the great deeds and victories of her heroes, in a sanctuary designed as a meeting-place for poets, could be under-

93 See DEGRASSI 1963: 463; Liv. 40.51–2.

94 The bracket in the Fasti Antiates maiores that appears to relate the entry to 21 December may, on the evidence of the later calendar, represent a 'slip of the pen'; a shift of one day on the occasion of a restoration is improbable.

95 Liv. 40.52.4–7: *Idem dedicavit aedem Larum permarinum in Campo. voverat eam annis undecim ante L. Aemilius Regillus navali proelio adversus praefectos regis Antiochi. (5) supra valvas templi tabula cum titulo hoc fixa est: 'Duello magno dirimendo, regibus subigendis, † caput patrandae pacis haec pugna exeunti L. Aemilio M. Aemilii filio auspicio imperio felicitate ductuque eius inter Ephesum Samu<m> Ch<i>umque, (6) inspectante eopse Antiocho, exercitu omni, equitatu elephantisque, classis regis Antiochi antea invicta fusa contusa fugataque est, ibique eo die naves longae cum omnibus sociis captae quadraginta duae. ea pugna pugnata rex Antiochus regnumque … eius rei ergo aedem Laribus permarinis vovit.' (7) eodem exemplo tabula in aede Iovis in Capitolio supra valvas fixa est.*

96 Enn. *Ann.* 156 Skutsch = 500 V.

stood only as an invitation to write histories. In the tradition of Hellenistic court poetry in Greek – but, in the absence of the social relationships of true courts, more in forms reflecting those between *patronus* and *cliens* – this represented a wide-ranging offer to poets, with no restrictions in respect of genre or material.[97]

The historicization of the calendar was accentuated – and limits to it set – by a further element: the lists of the highest Roman office-bearers, the consuls and censors. If the wall-painting of the Fasti Antiates maiores reproduces the *fasti* of the temple of Hercules Musarum as it existed at the time, omitting the more substantial caption, this might suggest that, in the year of dedication of the later calendar, a space was left free to the right of the calendar portion to receive further entries. The first such entry would perhaps have been the consuls and censors for 179: in any event, the two consuls for 173 would most probably have been already recorded. Symmetrically opposite to this portion, it is probable that there was a list section already offering a more comprehensive magistrate list:[98] 'Most likely, this was not a complete list of consuls from 509 or 507 or some other date posited as the beginning of the republic. To assemble such a record was a task for the rest of the second century and most of the first century BCE: a task that resulted in diverging lists and, ultimately, in the attempt at canonization by the Augustan Fasti Capitolini.'[99]

The size of the panel left for new entries in Fulvius' calendar painting is unknown. The relative proportions of the Fasti Antiates – 2.50 metres for the width of the calendar, 1.36 metres for three columns of magistrates – suggest three columns of text; but six columns, equalling the calendar in width, would also have been possible: the Fasti Antiates would then represent a snapshot of 67 BCE. Fulvius' original may have had space to accommodate a longer history, or even, at the cost of a loss of symmetry, been continued. A symmetrical arrangement, in from two to six columns, would have taken the history back 70 to 220 years. A six-column arrangement on each side would then lead to a year of inception at the time of the new foundation of Rome after the sacking by the Gauls (now thought to have occurred in 387): such an important event to Romans of Fulvius' time, and – as I will show – to Ennius, as their leading spirit.[100] Prior to the installation of the Fasti Capitolini, a list tailored in this way would lend monumental plausibility to Livy's assertion that records were lost owing to the Gaulish attack. But this is speculation: the exedra in the three-metre high podium of Fulvius' temple, as known from Imperial times, would have been some fifteen metres wide, and provided space for a whole variety of possible forms for a painting for which we have no basis of

[97] WEBER 1992; WEBER 1993; ZWIERLEIN 1982: 86–7.

[98] I am grateful to Denis FEENEY for privately and publicly insisting on this point (see 2007: 184–5). I try here to map out the consequences.

[99] RÜPKE 2006: 509–10; partly drawing on RÜPKE 1995. Cf. FEENEY 2007: 175–82 on the subversion of the Republican consular era by 1) AUC dating, by 2) the way the Fasti Capitolini deal with suffects, and by 3) the non-consular list of the *fasti triumphales*.

[100] See FEENEY 2007: 100–1 on Enn. *Ann.* 154–5 Skutsch.

measurement (such as that offered by the size of the letters in the Fasti Antiates).[101] We must remember, though, that it was the calendar, and not the lists, that gave the entire assemblage its name: *fasti*.

The lists of the individuals who bore responsibility for policy in the years in which they served, and above all the lists of consuls, provided precisely the element that the calendar lacked: names. Not until Caesar would this circumstance be changed.[102] The presence of these lists confirmed the historicity of the act of naming temples, simultaneously cited in the calendar, and so referred to the personal dimensions of a cult practice without infringing upon the sacral aura of the calendar text: on the contrary, the latter's religious character was given marked emphasis by the stream of names of goddesses and gods. In the converse sense, a normative dimension is discernible: magistrates were reminded of the impact of their actions on the sacral side of the *res publica*; form was given to an expectation or norm, and, at the same time, the prospect of reward, the reality of commemoration, held out to them. Fulvius' temple-construction or -extension project took place in a period that saw temple dedications almost every year: some fifteen in the previous twenty years.[103]

Had Fulvius' *fasti* been intended to provide historical dates, we would expect the dedication notices to have been couched in the form *XY YZ consulibus aedes dei/deae (a N. N.) dedicata est*. As such a longwinded entry would have made execution of the project appreciably harder, and the mass of text have risked rendering unrecognizable the basic framework derived from Cn. Flavius' *fasti*, drastic abbreviation would have been called for. Such an abbreviated rendering would then probably have ended in the nominative form of the god's name. But the Fasti Antiates and its successors unmistakably give the dative form.

How are we to understand the syntax of this dative form of the god's name in the *fasti*. Any association with the characters of the various days – the day is *fastus, nefastus, comitialis* to the god – must immediately be dismissed as absurd: the two elements are genetically unconnected. *Feriae alicuius* describes a special case not derivable from *NP*, but written out in full owing to its particular conceptual status. All attempts to explain the dative on the basis of constructions such as *dies est deo/deae proprius/sacer* must similarly be rejected: quite apart from the fact that we would expect the genitive in these cases, the theological conception underlying the grammatical construction is, with the exception of the *feriae*, unattested.

The second possibility is to see the dative as part of a cult instruction: 'On this day is due a sacrifice, a *supplicatio*, or the like, to …' We even have ancient testi-

101 The *exedra* is the most probable place for both the meeting of the *collegium poetarum* and the painting. 13 columns of the calendar would never have been distributed in the 12 niches of the podium as suggested by Guido Petruccioli in Haselberger 2008: 207.

102 Stressed by Feeney 2007: 184.

103 See the list in Latte 1960: 417.

mony for this interpretation: in the lavishly executed Fasti Vallenses, the dative of the god's name (regularly qualified by the name of a location, using *in* plus ablative) is frequently (when space permits?) complemented with *sacrificium publicum.*[104] This remains an isolated instance of such a formulation, however.

Nowhere in other *fasti*, even in comprehensive specimens such as the Fasti Amiterni, is a corresponding formulation with *sacrificium* to be found. Comparison with texts that perform the function of regulating a cult rapidly reveals the problem with this solution found by the Fasti Vallenses: having due regard for the spectrum of possible cult actions, we would expect precisely this variable to be specified; we would expect, for example, a distinction to be drawn between *supplicatio* und *immolatio*, as is found in the Feriale Cumanum.[105]

It is precisely the *feralia* that provide us with a deeper insight into the structure of the abbreviated entries. Thus in the Late Antique Feriale Campanum we find a text that mentions only the cult acts, with no clear polytheistic reference.[106] The Feriale Amerinum is difficult to assess, as only the line-ends survive, and we do not know the extent of the missing text.[107] What remains has gods' names in the dative, and guidance for sacrifices, with specification of the number of sacrificial animals: usually *h(ostia) V.*

The syntax becomes much clearer in the Augustan Feriale Cumanum. After the date at the beginning of the line, the structure follows a pattern such as: *Drusi Caesaris natalis. Supplicatio V[e]stae*; or: *Eo die Caesar togam virilem sumpsit. Supplicatio Spe[i] et Iuve[ntati].*[108] The particular event (the birthday of Drusus or the coming of age of Caesar (Augustus)) is either in the nominative style or in a complete sentence, and is followed by the cult instruction, and then the divinity in the dative. The event may also comprise a *dies natalis*, in which case the god's name occurs twice: *Eo die a[r]a Fortunae Reducis dedicatast, quae Caesar[e]m [ex transmari]nis provincis red[uxit]. Supplicatio [F]ortunae Reduci* ('On this day had been dedicated the altar of Fortuna the Homebringer, who brought Caesar back from the overseas provinces. Supplication to Fortuna the Homebringer').[109] Unfortunately, we have no second entry that might tell us whether the building itself was always associated with the genitive form of the divinity.

An entry in the Fasti Praenestini for 27 January has just such a genitive construction: *Aedi[s Castoris et Po]llucis dedicat[a est].*[110] However, a different formulation is much more frequent in these commentated *fasti*, the dative of the god's name in

[104] See *Inscr. It.* 13.2.149 on 5 and 8, as well as (without *publicum*) 21 and 27 August. Thus WISSOWA 1912: 474–5.

[105] *Inscr. It.* 13.2.279.

[106] *Inscr. It.* 13.2.283.

[107] *Inscr. It.* 13.2.281. The integration of the *V* at the beginning of the 5th line to form a date *[No]v.*, assuming only a small amount of space to the left of the extant portions, appears highly questionable, in view of the multi-lettered integrations of the beginnings of gods' names in the remaining lines.

[108] *Inscr. It.* 13.2.279, Z. 4–5.

[109] *Inscr. It.* 13.2.279, Z. 7–8.

[110] *Inscr. It.* 13.2.117.

the official text being continued with the clause *quod eo die aedis ei dedicata est* ('for on this day a temple has been dedicated to him/her'). But does this not suggest that the preceding dative should be understood as part of a cult instruction? Not at all: *Quod eo die* generally introduces explanations of brief calendar notes, being used especially in the explanation of *feriae* entries. It should thus not be translated '(a sacrifice) to the goddess, because she received a temple on this day', but '"to the goddess": (this stands here) because she received a temple on this day': this is the information a reader requires in order to understand the entry, and it is the information that Verrius Flaccus extracts from the *fasti*. Nowhere does he offer us an alternative explanation of such a use of the dative form of the god's name.

Examination of parallel texts has revealed the available spectrum of possible answers, without producing any definitive conclusion. The dative as a component of an entirely unspecified cult instruction has proved to be very problematic; but the alternative, the dative as part of an account of the foundation of a temple, has not been entirely substantiated. This alternative solution may, however, be underpinned by detailed perusal of the supplementary entries in the Fasti Antiates maiores.

Such entries using the dative arise not only on ordinary days, but also on *feriae*. The fact that they do not appear regularly in the latter case demonstrates that they are not merely amplifications of the abbreviations used for feast days. If, however, we look for criteria for their selective appearance, it becomes apparent that festivals where the divine recipient of the sacrifice is known are not singled out in this way. Thus, of the festivals where the condition of the date and the remainder of the line permits an assessment to be made, supplementary entries are absent from the Equirria, the Liberalia,[111] the Tubilustrium, the Fordicidia, in this regard also the Parilia qualified by *Roma condita*, the Robigalia, the Lemuria, the surviving Agonium, the Lucaria, the Neptunalia, the Consualia, the Meditrinalia, and, finally, the Divalia. The list is long enough to allow us to ensure that the common element is definitely the reason for these festivals' distinctive treatment: the divinities celebrated in the festivals' names, or commemorated by a cult in the course of the festival, had no real temples, or none that could have its foundation day on that day (Mars, Liber). In respect of the model for the Fasti Antiates, thus Fulvius' *fasti*, we are necessarily led to the conclusion that the dative must be understood in each case as a component – and shortened form – of the sentence *eo die alicui aedis dedicata est*.[112] This fits with the fact that there is no mention of important cult acts in these temples. The case of the Fasti Vallenses demonstrates that the above circumstance was not universally understood; we must wonder whether Fulvius did not, in the 'caption' to the epoch-making innovation represented by his *fasti*,

[111] On the problem of a temple of Liber, see DEGRASSI 1963: 425–6.

[112] This understanding is also shown by Ovid's *fasti* commentary, where the dative additions at the end of 1 January are explained (1.289–90, see RÜPKE 1994).

include some introductory words drawing attention to the *dies, quibus diebus aedes dedicatae sunt*.

This linguistically based interpretation of the dedication notices has confirmed our previous conclusions arrived at on historical grounds. It is further aided by the two notices that do not belong in the schema of the 'divine datives': these are *Roma cond(ita)* and *[Al]liens(is) die(s)*, associated with the Parilia, on 21 April and 18 July respectively. Both articulate unambiguously historical notices, belonging to the context of foundations and battles. The *dies Alliensis*, of course, referred to a day of misfortune; but the absence of any further references to such days – which defined all kinds of activities – indicates that the critical factor was not this particular character of the day, but the historically exceptional dimension of the Gaulish attack, whose date, like that of the foundation of Rome, thus appears to have become established in tradition by the beginning of the second century.[113] M. Fulvius Nobilior used Flavius' legal calendar purely as a foil for his own historical work: the above combination of circumstances allows of no other conceivable intention on his part. Even the brief references to games – *Loed(i) Apol(lini)* on 13 July and *(Ludi) M(agni)*, held from 8 to 10 September – do nothing to contradict this conclusion: just as in the case of temple-building, the context for such games was the investment of booty and the commemoration of victories.[114]

7.3 Ennius

The tradition leaves no room for doubt that there was a close association between the specific theological complexion of Fulvius' temple project and the Muses plundered from Ambracia. In view, however, of the great volume of artistic booty stemming from the Aetolian campaign, such an association is insufficient to establish a causal connection. Only Ennius' presence on the campaign, and his particular conception of the Muses,[115] turn a random collection of plundered items into a concept capable of architectural expression. If Ennius was central to the conception of the Museion as such, what role did he play in respect of the *fasti*, which, according to our analysis so far, formed an integral part of the building project?

113 RADKE 1990: 43. See Luc. 7.409: *Et damnata diu Romanis Allia fastis.*

114 Only brief mention can be made here of the problem of the inclusion of market days (*merkatus;* on the institution briefly DEGRASSI 1963: 377). They do not feature in the Fasti Antiates maiores, but appear in so many Imperial-Period calendars, beginning with the Fasti fratrum Arvalium, that we are entitled to assume their presence in Republican *fasti*. It is difficult to determine precisely when they were first introduced into the *fasti*. It is possible, on the one hand, to see them as appendages of the *ludi*, and thus as having come into the calendar in their wake; on the other hand, if we consider their economic function, we might conclude that indication of the 'annual markets' may have followed that of the 'weekly markets', implied in the nundinal letters, at any time.

115 Against MARTINA 1981: 68 see SKUTSCH 1985: 144–6, 649; ZWIERLEIN 1982: 86; SUERBAUM 1968: 43–295; *Arethusa* 39.3 (2006).

The bare outline that is all we are still able to discern of the historical conception expressed in Fulvius' 'caption' roughly corresponds to the position espoused by Ennius in the *Annales*. As a more precise picture of alternative contemporary thinking, and of the degree of general acceptance enjoyed by Ennius' position, is in most respects beyond our reach, intimations of correspondences between the *Annales* and Fulvius' *fasti* cannot be backed by any great weight of proof: parallel instances merely suggest the possibility that Ennius had a part in Fulvius' *fasti*; they do not prove it so.

Ennius knew Venus as the 'mother' of Romulus through the medium of Aeneas in the Trojan / Latin prehistory of the Romans.[116] Just as Fulvius' Romulus was the genealogically and politically inspired eponym of historical events, so Ennius' Romulus was the genealogically inspired eponym of topographical and politico-social entities, of Rome itself, and of the *tribus*.[117] Numa's calendar-related activities, which were roundly disputed in historical and antiquarian circles of the second century of the pre-Christian era,[118] find no reflection in extant Ennius fragments. The fact that the image of Numa as an organizer of the cult had wide currency nevertheless provides good reason for according him some attention.[119] Fulvius left behind no views in this regard, beyond those implied in the *dies natales* events. But his mention of the contemporary *lex Acilia* shows that his *fasti* extended the historical line into the present, as is provably also the case in respect of Ennius' work.

The common ground extends beyond the fundamental conception behind the historical account. We have direct testimony of Ennius' interest in the calendar, in the shape of a precisely dated note of an eclipse of the sun;[120] that interest is equally expressed in his assertion that the natural year has 366 days: reason enough to think that he played a role in the *fasti*.[121]

The Ennius connection might prove important in one further point. Our analysis of the introduction of the temple-foundation days into the *fasti*, and the way they combined with the magistrate lists, has shown that there was a historical perspective to this integration of elements alien to the *fasti*. The timeless character of the dedication days distributed across the year became transferred to the historical phenomenon of successive temple foundations, and the religious reading of those events as embodying the history of divine epiphanies and cult diversification acquired a political dimension. In the history of ideas, this kind of inversion is referred to as euhemerism.

Ennius' *Euhemerus* is not merely eponymous for the Latin tradition. In this treatise, he concerned himself directly with the origins of temples and – in this order

116 Enn. *Ann.* 58 Skutsch = 52 V.

117 Enn. *Ann.* 77 Skutsch = 82 V; *Ann.* test. lix Skutsch = Varro, *Ling.* 5.55, cf. SKUTSCH 1985: 252–3.

118 See Macrob. *Sat.* 1.13.20; Censorinus, *DN* 20.4.

119 Enn. *Ann.* 114–18 Skutsch = 120–4 V, especially the list of *flamines* 116–18 Skutsch = 122–4 V.

120 Enn. *Ann.* 153 Skutsch = 163 V. See RÜPKE 1995a: Ch. 4.1. SKUTSCH too (1985: 313) notes the association.

121 'Speculations on the calendar are attested for him … He probably assisted Fulvius in annotating the *fasti*.' Thus SKUTSCH (1985: 314), with reference to Enn. *Inc.* 32 (from Censorinus, *DN* 19.2).

– annual festivals. Ennius followed a Greek model in his work, and the fragments that survive contain no explicit ancient history of *dies natales templorum* in the city of Rome. It is difficult, today, to arrive at an appropriate assessment of his position. To us, such a critique appears downright radical. It may, however, not even have occurred to Ennius and his readers that an explanation of the genesis of the gods and the cult necessarily implied atheism and denial, not to mention a call for action in the real world: this would certainly have been the last thing the temple's patron needed.[122] Despite its critical implications, even for readers in the ancient world,[123] a philosophical treatise such as *Euhemerus* constituted less an attack on religion, than the broadening of a religious conception centred on the cult into one that also encompassed theorizing in prose form.[124]

Fulvius' *fasti* demonstrate that, with Ennius and the narrow circle of which he was a shining light, for the very first time in Rome a level of intellectual sophistication was reached sufficient to lend plausibility to the intended agenda reconstructed here for the Museion *fasti*. In view of the degree of closeness between the two men, it appears to me in no way excessive to categorize Ennius as Fulvius' intellectual collaborator in the creation of the *fasti*.

The parallel with Ennius' *Annales* is twofold. In terms of content, the impulse to create a record is also found in the mentions of consul pairs and iterations in Ennius' poetry.[125] In terms of form, in both contexts we find an interest in continuing a work that was, in principle, already concluded: just as, in the calendar, the final date was the *dies natalis* of the Camenae, so the *Annales* too, in their original conception, should probably have closed with the bringing of the Muses to Rome by means of Fulvius' temple,[126] but were then continued by Ennius until his death. Book 16 may mark the beginning of this process of continuation, but, when Ennius opens the book with a reflection on the vain attempts of great men to create an eternal name (*nomen*) for themselves with statues and magnificent tombs, he has in mind the intention of Fulvius' consul list within the writers' place of assembly.[127] Only the poet can create true fame, as Ennius writes in his *Scipio*: *Nam tibi munimenta mei peperere labores*.[128] For Fulvius and Ennius, the calendar in the Museion is the continuation of the *fasti* by other means.

122 The *Euhemerus* was written between the early *Scipio* (see SCHOLZ 1984: 195) and the beginning of the *Annales*, in the second half of the 180s (see SKUTSCH 1985: 3–6).

123 See Cic. *Nat. D.* 1.119; WINIARCZYK 1994.

124 See BURKERT 1985: 305 (see also 332–7) for the rise of philosophy in Greece, and SCHEID 1992; FEENEY 1998; RÜPKE 2009.

125 Enn. *Ann.* 290 Skutsch = 295 V: *Quintus pater quartum fit consul.* 304–6 Skutsch = 303–5 V: *Additur orator Cornelius suaviloquenti / ore Cethegus Marcus Tuditano collega / Marci filius.*

126 SKUTSCH 1985: 570; see also 649. LEO (1913: 170) probably sets the end too early with Nobilior's campaign.

127 Enn. *Ann.* 404–6 Skutsch = 411–3 V; cf. SKUTSCH 1985: 567–9; ZWIERLEIN 1982: 97–102; see also SUERBAUM 1968: 165–239.

128 Enn. frg. *Var.* 7; see also 1–2 V.

7.4 All *fasti* are Fulvian *fasti*

We have thus given substance to the form taken by Fulvius' *fasti*; they comprised calendar and consul list, gave each a particular form, and, incidentally, bestowed the name *fasti* on the combination thus created; so that, now, it appears that to speak of the influence this wall-painting exerted on the *fasti* in Antium would not amount to a circular argument. The Fasti Antiates follow the Fulvian pattern down to the last detail. But this also largely applies to the Fasti Arvalium, the first Augustan marble *fasti*, which can themselves be seen as a prototype (but not necessarily a direct model) for the other Imperial-Period inscriptions in marble. As the possibility that these later calendars took the Fasti Antiates as their model must be excluded, we are entitled to treat the Fasti Fulvii as the archetype of all Late-Republican *fasti* in whatever medium.

It is highly unlikely that it was the original intention of the creator or commissioner of the *fasti* in the temple of the Muses to become the progenitor of a line of Late-Republican specimens. Not only the mode of execution, but above all the place of 'publication' points to the effect he had in mind. Whatever the terminological distinction between *scribae* and *poetae* prevailing at the time the temple was built, Fulvius envisaged an audience of poets, not one of copyists. Despite the fact that the *fasti* consisted largely of letters of the alphabet, they constituted a wall-painting, whose model must thus be sought in the genre of triumphal painting.[129] An ancestor, M. Fulvius Flaccus, is mentioned as a commissioner of such works.[130]

The new medium of self-portrayal was doubtless seen as a boon by the nobility.[131] And we must not forget the world of antiquarian literature. It was most particularly Fulvius' *fasti* that provided both text and impetus for the first commentary on the calendar, that of Iunius Gracchanus. He was then emulated by Varro, in the relevant passages of *Antiquitates rerum humanarum et divinarum*, the standard work, which then influenced both Ovid's and Verrius' commentary on the *fasti*, and into Late Antiquity remained a stimulus to other writers.[132]

[129] See LEHMANN-HARTLEBEN 1926: 123–4; HÖLSCHER 1980: 269; HANNESTAD 1986: 36–7. On the audience for this kind of prestige art see HÖLSCHER 1984.

[130] Festus, *Gloss. Lat.* 228.20–3 L.

[131] It may even have been owing to Fulvius' combination of eponymous list (as core element) and calendar that, two decades later in 153 BCE, the date of investment of consuls was moved to 1 January, thus bringing the eponymous year and the calendar year into step.

[132] An influence on the *Annales* of L. Accius, which, judging by the longest surviving fragment (*Carm.* frg. 3 from Macrob. *Sat.* 1.7.36), also discussed the (pre)history of Roman festivals, in this case the Saturnalia, can only be guessed at. To see the work as a whole as 'a depiction of the annual festivals' (SCHANZ AND HOSIUS 1927: 135) is, to say the least, not a safe conclusion to draw from the five fragments.

8

From Republic to Empire

8.1 Caesar's Calendar Reform[1]

With Caesar, we enter a period that, in the Fasti Antiates maiores, produced the first extant calendar. More important still than the historical setting of this isolated wall-painting, however, is the character of the time as marking the prelude to the proliferation of *fasti* in the Augustan period. From this perspective, two questions demand our particular attention. What significance should be assigned to the technical reform that turned the 'Republican' (or 'Numan') into the 'Julian' calendar? And what place did the *fasti* occupy in the general consciousness, or rather, considering the nature of most of our sources, the consciousness of the elite?

Measured against its intended correspondence with the natural year, the Roman calendar was in considerable disarray in the 50s.[2] In this period when political discourse was becoming enormously more heated, and in different areas the use of religion as a political weapon had been carried to the extreme (the name P. Clodius comes to mind), increased recourse was no doubt also had to the device of prolonging office by intercalating, or by refraining from doing so. As was the case with other obstructive ploys used, however, no really fundamental criticism was expressed, even here: the aim was to master the tools on offer, not to abolish them.[3] Of contemporary sources, it is only Cicero's draft constitution, in *De legibus*, that offers some mild criticism. Cicero as legislator would wish to oblige the *pontifices* to keep scrupulously to Numa's rules for intercalation.[4] However, his

[1] Plin. *HN* 18.211–12; Suet. *Iul.* 40; Censorinus, *DN* 20.6–11; Dio Cass. 43.26; Macrob. *Sat.* 1.14; MALITZ 1987.

[2] DRUMANN AND GROEBE 1906: 753–827; BEAUJEU 1976; BRIND'AMOUR 1983: 40; problematical: RADKE 1990 (see BRISCOE 1991 and RÜPKE 1992b).

[3] MALITZ 1987: 106–7.

[4] Cic. *Leg.* 2.29: *Quod* <ad> *tempus ut sacrificiorum libamenta serventur, fetusque pecorum quae dicta in lege sunt, diligenter habenda ratio intercalandi est ...*

The Roman Calendar from Numa to Constantine: Time, History and the Fasti, First Edition. Jörg Rüpke.

intention at the level of sacral law was not to bind feast days to astronomical data, but to facilitate the detailed observance of the rules for sacrifices – animals of a particular age, precisely defined fruits (of the field) – by correlating the festal calendar with the seasons. Cicero did not have wholesale reform in mind; he wished to curb excesses in current practice.

Cicero's work reflects the situation in 52 BCE. With his suppression of the revolt of Vercingetorix in this year, Caesar had largely concluded his conquest of Gaul, thus, by the beginning of 49, establishing the basis for embarking upon civil war against Pompey and the majority of the Senate. With the victory at Pharsalus, and the assassination of Pompey in Egypt, the course of events was already largely decided. It is only at this point, after Caesar's return from the East[5] in October 47, that we find evidence of his thoughts turning to reform. Fully intent on obtaining a third term as consul, Caesar made use of his control over the electoral assemblies to fill magisterial and priestly positions at his discretion, at the same time expanding their number, and, once again in Rome's history, attempting to come to grips with the debt problem.[6]

The sources are silent as to what political motivation underlay the reform, over and above considerations of technical and scientific advance[7] in respect of the calendar. The circumstances under which the reform began, extensions of the administrative apparatus, and adjustments to the electoral rules, reveal the beginnings of a more comprehensive process. The regular proceedings that governed Caesar's own consular year, with its seasonally determined, calendar-based beginning in autumn, may have made him vividly aware of the urgent need for changes.

That we may interpret the reform as deriving from Caesar's person and intentions is demonstrated by the procedure adopted for it. The new calendar was developed by a commission appointed by Caesar, and, most importantly, made up of non-Roman experts;[8] prominent among these was Sosigenes, who evidently had overall charge of reform proposals.[9] There is no evidence for the participation of the pontifical college, entirely dominated by Caesar after the elections of 47 and 46,[10] except for a *scriba* named M. Flavius who assisted Caesar, and in whom may once again be concealed a *pontifex minor*.[11] This would be yet another case of the prominent participation of *pontifices minores* in calendar-related affairs, based on their original function in the empirical determination of the Nones, but finding particular expression in their role in major reform projects, perhaps as a seal of quality for the eyes of the public. Cn. Flavius with his *fasti* would be the first instance, Cn. Terentius in 181 BCE with his discovery of the *libri Numae*

[5] MALITZ 1987: 115, 112.

[6] MEIER 1995: 414–20.

[7] Thus MALITZ 1987: 111–12, 117.

[8] Plut. *Caes.* 59.5; Dio 43.26.2.

[9] See Plin. *HN* 18.211–12.

[10] See RÜPKE 2008.

[11] RÜPKE 2008, no. 1655.

(rapidly condemned as undesirable and burned) the second, this time ill-famed,[12] and M. Flavius now the third. Implementation of the reform finally depended on an edict of the *dictator iterum* in the second half of 46;[13] the Senate was fully informed, but did not participate in the decision-making process.[14]

A brief analysis of the reform's content will make it easier to understand its legal structure. From the technical/astronomical point of view, there were three aspects to the reform. Most significant in terms of its reception was the redefinition of the length of the natural year at 365.25 days, reflected in a 'civic', 365-day year, and four-yearly intercalation. Historically, this outcome had already been achieved by the Metonic cycle and comparable systems;[15] in Rome too, an approximation of 366.25 days had already been obtained by the decemviral reform at the end of the fourth century BCE.

More importance therefore appears to attach to the second aspect already addressed, the implementation of the calculated year-length in a system that minimized the extent and frequency of intercalations. Instead of twenty-two intercalary days in a two-year cycle (or entire intercalary months every two or three years in the Greek world), Rome now introduced a calendar in which a single day had to be intercalated every four years, in an entirely regular pattern. This is the element that rendered the calendar, that is to say the intercalation process, immune from any political or financial exploitation and associated attempts at manipulation.[16] This is the element that removed the thirteenth month from all written calendars of the time, thus changing their appearance markedly. And, finally, this is the element that much diminished the religious role of the *rex sacrorum*, the priest who implemented ritual intercalation.

The third aspect is of little import in comparison, although it may have persisted the most powerfully in the memories of Caesar's contemporaries. In order to bring the calendar dates into step with their analogues in nature (in which state they were then to remain),[17] a great number of days had to be inserted: over and above the normal twenty-three day intercalation at the end of February, sixty-seven more days in all, distributed between a *mensis interkalaris prior* and *posterior*.[18] If we compare this prodigious intervention with the task undertaken by the *lex Acilia*, to correct the then current discrepancy by gradual intercalations, it again becomes clear that the utmost possible elimination of intercalation, with all its

[12] RÜPKE 2008, no. 3222.

[13] Macrob. *Sat.* 1.14.13.

[14] MALITZ 1987: 115–16 with reference to Cass. Dio 43.26.1–27.1.

[15] ZION WACHOLDER AND WEISBERG (1971: 240–1) show that both the Metonic and the contemporaneous Palestinian 19-year cycles originated in Mesopotamia. See also OPPENHEIM 1969: 125; GALIL 1991: 372. No systematic intercalation is attestable in Babylonia before the fourth century BCE; cf. SAMUEL 1972: 21.

[16] Censorinus, *DN* 20.7.

[17] MALITZ 1987: 123.

[18] Censorinus, *DN* 20.8: *duos menses intercalarios dierum LXVII in mensem Novembrem et Decembrem interponeret, cum iam mense Februario dies III et XX intercalasset, faceretque eum annum dierum CCCCXLV.* Cass. Dio 43.26.1; on the dating: Cic. *Fam.* 6.14.2.

uncertainties and irregularities (to quote Macrobius' impudent phrase the main characteristic of the Republican *anni confusionis*),[19] was Caesar's central purpose.[20]

These three steps enabled the calendar to be maintained in durable, constant and precise synchrony with the solar year; the linking of the 'civil' 25 March with the spring equinox was pivotal. In a civilization that had learned always to maintain the astronomical calendar as a second system, based on calculation, observation and parapegmata, and independent of the civil calendar, this led to a long-term cognitive reorientation. For the first time, astronomical events, and meteorological events in association, could be described on the basis of civil dates, something that Cicero had observed in 70 BCE as an alien phenomenon, characteristic of Sicily and the Greek world.[21] That, according to the testimony of the weather calendar passed down by the elder Pliny (*HN* 18.207ff.), Caesar concerned himself with the dates of rising and setting of stars shows that he was aware of the gradual inception of a conceptual change. Although Caesar's immediate contemporary Cicero mockingly remarked that even the stars now moved to order,[22] this very conceptual reversal, this naturalization of the calendar, may have been highly attractive. Ovid, in his commentary on the calendar, revels in the possibilities thus opened up.[23]

Notwithstanding the 445 days of the *annus confusionis ultimus*, 46 BCE, the process of recasting the year's shape was marked by the utmost prudence. The opportunity to subject Caesar's way of proceeding to precise scrutiny by means of a detailed reconstruction is all the more valuable in that such circumspection is more in keeping with our image of Caesar's heir than with our normal idea of the late Caesar himself.[24] The names of the months remained unaffected, and the calendar acquired no new *feriae*; such changes as were made in this regard belong to the two following years.[25] The time of year for carrying out intercalations remained unchanged.[26] The sources do not tell us what changes ensued to the function of the *rex sacrorum*.

The reduced length of the intercalary period meant that the annual deficit of ten days formerly covered by the *mensis interkalaris* had to be distributed between the remaining twelve months. We learn from the comprehensive reference in Macrobius that Caesar's guiding principle was the preservation of the internal monthly structure wherever possible, and that the considerable care he devoted to this task of redistribution was matched by the scrupulous way in which his

19 Macrob. *Sat.* 1.14.3. Misunderstood by MALITZ 1987: 119 n. 81.

20 See Censorinus, *DN* 20.8: *providens in futurum, ne iterum erraretur; nam intercalario mense sublato annum civilem ad solis cursum formavit.*

21 Cic. *Verr.* 2.2.129; FEENEY 2007: 196.

22 Plut. *Caes.* 59.3; HOLLEMAN 1978.

23 RÜPKE 1996; FEENEY 2007: 196–206; PFAFF-REYDELLET 2009; RÜPKE 2009b.

24 In general, ZECCHINI 2001.

25 See Ch. 6.2.

26 Censorinus, *DN* 20.10. See RÜPKE 1995a: Ch. 6.1.

measures were observed.[27] The complex month of February was left entirely untouched, as were the four months of March, May, Quintilis, and October, which already had the maximum thirty-one days. The remaining months were lengthened by one or two days, inserted in each case immediately before the last day, so that the position of the Nones and Ides remained unchanged relative to the beginning of the month.

The additional days at the ends of these months, that is to say after all the month's feast days, resulted, for all dates after the Ides, in an increased interval until the next Kalends, calculated backwards from the Kalends. The fact that Caesar accepted this alteration shows that, in respect of the festivals of the second half of the month, the critical interval was seen to be the interval from the Ides, which by this means remained the same, and not the interval to the month's end. But the forward reckoning involved here appears to have been more than simply implicit.[28] The mode of insertion was explicitly chosen in such a way that the announcement of *feriae* remained unaffected.[29] The *indictio* referred to by Macrobius can relate only to the monthly announcement of festivals by the *rex sacrorum*: taken together with Varro's use of the present tense in describing this institution,[30] here is further, and until now overlooked testimony to its survival, at least beyond the end of the Republic. At the same time, a dissociation between the sacral and political/juridical calendars is indicated, extending even to the mode of counting, and showing the Ides of each month to have been the religious reference point for the entire month.[31]

The choice of character for the new days confirms the political and religious interpretation presented in earlier chapters. By categorizing these days as *dies fasti*, Caesar avoided extending the scope for political activity (statistically, characterization as *C(omitia)* would have been the norm), while feeling no particular necessity for deploying the further degree of restriction represented by *N(efas)*.[32] Comparison of the pre-Julian parapegm with Augustan calendars essentially confirms the antiquarian narrative. There are clear differences only in three cases in which the daily character *EN* changes to *F* (on 16 October) or *NP* (on 15 and 17

[27] See Macrob. *Sat.* 1.14.6–12.

[28] SUERBAUM (1980: 330) goes so far. See also Ch. 8.1.

[29] Macrob. *Sat.* 1.14.8–9: *quintanas tamen habent nonas, et ab idibus illis sequentes kalendae in undevicesimum revertuntur, quia Caesar quos addidit dies neque ante nonas neque ante idus inserere voluit, ne nonarum aut iduum religionem, quae stato erat die, novella comperendinatione corrumperet. (9) sed nec post idus mox voluit inserere, ne* feriarum *quarumque violaretur* indictio, *sed peractis cuiusque mensis feriis locum diebus advenis fecit.* See also 1.15.8: *quia peractis totius mensis feriis dies suos rei divinae cautus inseruit.* Corresponding factually to Censorinus, *DN* 20.9: *Eosque dies extremis partibus mensum adposuit, ne scilicet religiones sui cuiusque mensis a loco summoverentur.*

[30] Varro, *Ling.* 6.13.

[31] Thus I cannot accept Feeney's treatment of the 'anniversary', and his interpretation of the handling of birthdays after the reform (2007: 148–60).

[32] Macrob. *Sat.* 1.14.12: *Nam ideo novos dies circa finem cuiusque mensis inseruit, ubi finem omnium quae in mense erant repperit feriarum, adiectosque omnes a se dies fastos notavit, ut maiorem daret actionibus libertatem, et non solum nullum nefastum sed nec comitialem quemquam de adiectis diebus instituit, ne ambitionem magistratuum augeret adiectio.*

December, i.e. the Consualia and the Saturnalia), while the Opalia (19 December) are marked *N* (pre-Julian) and *NP* (Augustan). All four distinct cases represent changes that can have been motivated only by changed sacral evaluations of the days in question: the notation in the calendar follows developments; it does not determine them, if we set aside Flavius' treatment of the *nundinae*, which was soon corrected by the *lex Hortensia*.

This can be shown more precisely, at least for the two December dates. The calendar does not itself undertake re-evaluations: it very belatedly takes account of religious developments and changes in pontifical conceptions. Caesar uses major changes and necessary innovations caused by factors external to his project, in order to carry out an overdue *aggiornamento* of religious practice.[33] In the case of the Saturnalia, developments that raised their status as a festival at the end of the third century are mentioned explicitly in the sources.[34] In the context of this multi-day festival, comprising the two previous and two subsequent days, the two other feast days, in a certain manner doublets of the Consualia and Opiconsivia on 21 and 25 August (or Sextilis), appear also to have been promoted to actual *feriae* status in the course of the enhanced Saturnalia celebrations. In contrast, the sacrifice on 16 October, the day following the October Ides, with their ancient rituals of the *ludi Capitolini* and the *October equus*,[35] may have undergone a process of devaluation.

The great care Caesar took in respect of existing religious institutions when he was actually configuring the reform invites comparison with the French Revolutionary Calendar.[36] The calendar promulgated on 24 October 1793 – from then on 3 Brumaire of year 2 of the Republic – broke with the traditional Gregorian calendar at every possible point: beginning with the names of the months,[37] continuing with their length (twelve to thirty days), and the names of the days and seasons, and extending to a decimal system for subdividing the day into ten decimal hours of one hundred decimal minutes. Despite certain simplifications in the lengths of months, compensated for, on the Egyptian pattern, with five epagomenic days at the end of the year, no chronographic advance was achieved over the old calendar.[38] The break itself formed part of the purpose that the calendar's ahistoric, but natural, rational, and secular image should serve as a symbol and model for the new society that was to be created.[39]

More than the new elements of the national cult,[40] in France it was the gradually evolved rules governing the *décadi*, designed as a replacement for Sunday, that

[33] Against WISSOWA 1923: 381–2 and MICHELS 1967: 186. I follow DEGRASSI 1963: 334.

[34] Festus, *Gloss. Lat.* 432.9–11 L; Macrob. *Sat.* 1.7.24; cf. Varro *ap.* Macrob. *Sat.* 1.8.1.; Liv. 2.21.2; 22.1.19–20.

[35] RÜPKE 2009a.

[36] MEINZER 1992.

[37] NADJO 1991.

[38] MEINZER 1992: 40–2.

[39] ZERUBAVEL 1977; BACZKO 1984; MEINZER 1988: 25; OZOUF 1975, 1976, and BAXMANN 1989.

[40] LEFEBVRE 1957: 364; MEINZER 1992: 54–7.

were to show how deep the intervention meant to penetrate.[41] But it was on this very point, which had the most enormous impact of all on the rhythm of life, that the reform came to grief, on the land in particular, and despite isolated but considerable successes;[42] it was precisely in the contrast experienced after the abolition of the Revolutionary Calendar that the old weekly rhythm acquired a power it had never had before.[43]

The symbolic overkill that underlay the failure of the French Revolutionary Calendar[44] teaches us how properly to evaluate the success of the calendar reform that founded our own calendar down to the utmost detail. Just as had occurred in Cn. Flavius' project, substantial measures were taken, while at the same time the greatest possible effort was made to preserve the calendar's outward appearance. Even this achievement, however, would not have been possible without the enormous institutional changes of the late fourth century, and Caesar's revolutionary dictatorship.

Against the background of the dates uncovered by our examination of the content of the reform, we can now turn again to the authority by whose function Caesar set it in train. The list prepared by Caesar's *pontifex minor* shows that, for his changes to the characters of days, the dictator may have called pontifical opinions in aid; we do not know of any formal resolutions, just a series of concrete, individual decisions, year upon year. All the other changes were unproblematic from a sacral point of view: only the length of the intercalary month and the shape and length of the final part of the month were subject to changes. Thus the timeframe affected was either non-existent in terms of civil and sacral law,[45] or a space free of festivals: no religious authority was called for here.

Only the realization of the intercalation programme presented problems. How was the *rex sacrorum* to be obliged to cooperate? This barrier too was relatively easy to overcome. It was in principle possible to force a priest to carry out sacral functions, either by means of new institutions – quite a common expedient – or, in individual cases, by legal pressure. That it required a resolution of the people to bring the *pontifex maximus* Cn. Cornelius Scipio Barbatus to participate in the dedication of the Temple of Concordia in 304 BCE may have been unusual,[46] but it was not illegitimate. The *lex Acilia* had prolonged the intercalation procedure by the involvement of the *pontifices*, but, just as it was possible then to charge the *pontifices* in general with management of the process, so it was possible now to call upon them by the statutory means of the *edictum* to have the *rex sacrorum* perform an intercalation every four years. The fact that Caesar himself held the office of *pontifex maximus* may have made the reform more acceptable: the placatory

[41] Mathiez 1904; Aulard 1927: 152–5; Sydenham 1974: 39, 154; Meinzer 1992.
[42] Meinzer 1988: 27–59, 1992: 77ff.
[43] Meinzer 1988: 60, 1992: 150–6.
[44] Zerubavel 1977.
[45] See Rüpke 1995a: Ch. 6.1.
[46] Liv. 9.46.6.

involvement of a *pontifex minor* attests to Caesar's awareness of the problem. 'Constitutionally', neither factor was necessary.[47] The question also played no part in Augustus' later corrective measure.

We now come to the main institutional problem, the implementation of the new system. The arithmetic of the Julian year required one intercalary day every four years, to compensate for the annual deficit of one quarter of a day in the civil year as opposed to the solar year. But the *pontifices* began to intercalate every three years: incidentally, an important indication that ritual intercalation by oral announcement of the *rex sacrorum* continued to exercise its communicative function, and had not been replaced by an intercalation plan circulated in written form. Was this stupidity, or malice on the part of the priestly college?

We may exclude the second alternative; in whatever year we choose to place the first intercalation intended to follow the new rules, the pontifical college was dominated by Caesareans, and had no conceivable interest in a virtually imperceptible act of sabotage. Any attempt at protest would have inevitably resulted in the restoration of the *status quo ante*. Ignorance can also be rejected, in view of the fact that the reform was supported by accompanying academic treatises, such as *De astris*, a work ascribed to Caesar.

Macrobius' exhaustive account provides one key to the problem's solution.[48] This detailed account may go back via Suetonius to contemporary sources, and even, in central points, retain phrases from official edicts. As we have excluded 'stupidity', simple misunderstanding of an intercalation *quarto anno* (scil. *confecto*) as *quarto anno, id est tertio anno confecto* cannot be a sufficient explanation. The problem does not lie in the mathematics, but in a practice compatible with the wording of Caesar's edict. According to Macrobius, this demanded intercalation after completion of the fourth year, before the beginning of the fifth; thus at the end of the fourth year. But such an intercalation was impossible in Rome, and still is for us: February is when we intercalate. Intercalation after completion of the fourth year cannot occur until the next February, that is to say *after* the beginning of the fifth year. In this dilemma, the only available resort was to interpret *conficere* in an ingressive sense: 'after the fourth year had been ushered in'; intercalation occurs *anno ... quarto non peracto sed incipiente.* Not stupidity then, but *sacrificium intellectus*. Caesar's rules were followed to the letter, against better knowledge of their intent: nothing strange about that, for people schooled in the law.

This is the very point Augustus dealt with in formulating his directive, which occurred in 8 BCE, at the same time as the Senate resolution renaming the month of Sextilis as Augustus, and perhaps in one and the same statutory context: intercalation was to occur *quinto quoque incipiente anno*, of course after completion of the fourth year, but after, not before, the beginning of the fifth; he thus had no need to fear that the fifth year would then be passed over as starting year.

47 An alternative view in SEGUIN 1988: 417.

48 Macrob. *Sat.* 1.14.13–15.

The practical problems go back to a juncture at which Caesar could nevertheless have acted primarily as *pontifex maximus*, and not as dictator. This did not concern the calendar reform, but its preparation by the extreme prolongation of the year 46: a necessary element, but one that must be distinguished from the actual recasting of the calendar.[49] The insertion of two additional intercalary months after the regular intercalation in February[50] took place between November and December.[51] There was no precedent for such an action, and its practical implementation had to be carefully considered. As with regular intercalations, even after the reform, it could not be announced in writing, but had to be promulgated in a concrete situation, by word of mouth. To remind ourselves of the regular February intercalation: it was tied to an appearance of the *rex sacrorum* and to a particular day, the Regifugium, that could be reinterpreted without difficulty. The year offered no second such opportunity, but analogous occasions could be sought.

When did the intercalation occur? Censorinus' words, *in mensem Novembrem et Decembrem interponeret*, in avoiding the word *inter*, tell us that the point chosen for intercalation did not lie precisely between the months: if it had, it would have entailed an alteration to the character of the following December Kalends. The additional days in the reform calendar give a clue to the point of time actually chosen. As *dies fasti* among the usual, ancient *dies comitiales* at the end of the month, they are easy to identify; Macrobius lists them individually, and Verrius Flaccus in his Fasti Praenestini highlights these days with legends such as *hunc diem divus Caesar addidit*.[52] They are usually the second to last, sometimes third to last day of the month; only in April, where the *ludi Florae* stretch from 28 April onwards, past the end of the month and into May, is the fifth to last day added. Evidently, even where no concrete ritual link existed with the following Kalends, the intention was to avoid displacing the old *dies pridie kalendas* (it was, perhaps, on the evening of this day that the *pontifex minor* set off make his observation of the new moon, the result of which was announced the next morning, on the Kalends). It follows that appropriate points for an intercalation in November/ December of 46 BCE would have been 29 November and 29 December, both days being free of their own cult events.

And how was it carried out? My hypothesis is that, on the last day of November, the *rex sacrorum* appeared in the Comitium and announced that day as *ante diem VI kalendae interkalares priores*. The number evoked associations with the Regifugium in February; the practice of announcing *kalendae interkalares* not for the moment of intercalation, but for a subsequent day, was not only age-old, but also had good precedents in the Republican calendar in years in which twenty-three days were added. At the same time, this manner of announcing the date

[49] It was so perceived at the time: while Censorinus (20.8) and Macrobius (1.14.2–3) cover the intercalation of two additional months before going on to describe the reform, the opposite is the case in Suetonius (*Iul.* 40.2).

[50] See the assessment of the sources in BEAUJEU 1976: 24.

[51] Censorinus, *DN* 20.8.

[52] Macrob. *Sat.* 1.14.9; *Inscr. It.* 13.2.117. 131. On the following see Ch. 3.2.

solved the problem of the length of the period to be intercalated. After the five-day 'prelude', it only remained for the two intercalary months to be inaugurated on the Kalends by a *pontifex minor* stationed in the Curia Calabra, using the formula *septies te kalo, Iuno Covella*; in this way, the two months of thirty-one days each, amounting overall to an intercalated period of sixty-seven days, could be promulgated in a way that did not offend ritual, and at the same time was technically unambiguous: there is a one-to-one correspondence between *nonae septimanae* and months of thirty-one days in the Julian Calendar. The rules required that the thirty-first day of the *mensis interkalaris posterior* was followed by the last day of November, redefined at the outset; December could then run its normal course.

Where does this reconstruction take us? It is an attempt to find a solution to a problem to do with astronomy and the calendar, in the context of given principles of sacral law and ritual procedures. Fortunately, the hypothesis has explanatory potential; otherwise, the effort would be of no avail.

Knowing that intercalation took place in November or December, one wonders whether the Fasti Praenestini, which takes such careful note of the added days, might not also contain some clue in respect of the exceptional intercalation that prepared the reform. We do in fact have residual traces of such a clue. To the left of the extant entry for 31 December, we can discern at the right-hand edge of the column for November, the rest of which has been lost, an entry that, based on the normal structure of the *fasti*, must relate to 30 November (Julian), or 29 November (pre-Julian). Still discernible is only the close of the first, evidently long, line, which ends with *[–] COS*, thus with a date.[53] Unfortunately, 30 November survives only in the rather sparsely annotated Fasti Maffeiani, where the still pre-Julian note *C* indicates no peculiarities.[54] No literary or other epigraphic sources exist that might allow us to assign what was nevertheless an important event to this date.[55] Reconstruction of the intercalation process, with the aid of Censorinus' text, provides the basis for integrating this textual remnant: *[Hoc die II menses interk(alares) dierum LXVII interkalat(i) sunt C. Caesare III M. Aemilio] co(n)s(ulibus).*

A second positive result concerns our understanding of a singular date in a letter from Cicero to Q. Ligarius, written immediately after a supplicatory visit to Caesar *a. d. V. K. intercalaris priores*.[56] Such dates before the intercalary Kalends never occur elsewhere in contemporary accounts: they could not, as intercalation was announced on the previous day (*pridie*) at the earliest.[57] The fact that Cicero can already speak of *kalendae intercalaris priores* here reinforces the epigraphic evidence that the entire block of time was inserted at one time.

Thirdly, we are provided with an answer to our initial question as to the functions under whose authority Caesar undertook the reform. The position from

[53] *Inscr. It.* 13.2.135; illustration: 138.

[54] *Inscr. It.* 13.2.82; Fasti Antiates maiores: *Inscr. It.* 13.2.23.

[55] Degrassi 1963: 533: '*Quam ad rem adnotatio pertinuerit, non liquet.*'

[56] Cic. *Fam.* 6.14.2.

[57] Such dates could of course be formulated retrospectively: see Cic. *Quinct.* 79 from 81 BCE.

which Caesar is most likely to have been able to obtain, or perhaps compel, the cooperation of the *rex sacrorum*[58] and a *pontifex minor* in our reconstructed intercalation is that of *pontifex maximus*. It is precisely for this step, which merely prepared the way for the reform, that Censorinus describes Caesar as *pontifex maximus*; only for the actual reform, not for this preliminary stage, does Macrobius use the expression *edictum*, which refers to the dictator as instigator of the new calendar. Finally, it is for this same preliminary stage that Macrobius attests to the employment of the *scriba*, probably the *pontifex minor* M. Flavius. Caesar defined and inaugurated the Julian Year as dictator. It was Caesar the *pontifex maximus* who ensured that the pre-Julian 1 January 45 could be used as the Julian 1 January.

Once implemented, the calendar was used without discernible enthusiasm, but also without any real resistance. The population may have coped with the two preparative intercalary months in the habitual fashion, by treating them *pro momento temporis*:[59] we know of no problems arising. Caesar was able to make good use of the additional time in his third consulate for reforms, and to prepare the Spanish campaign. On the other hand, difficulties were experienced in the second half of the lengthened months in the matter of fixed dates, that is to say birthdays. The established principle in sacral law, that the position of feast days was not tied to the vernacular 'date', that is to say the interval to the following Kalends, but to the interval from the preceding Ides, came up against the weight attached in the private realm to that same vernacular dating system.

The Saturnalia gave rise to more such irritations. Noted in the pre-Julian calendar *a. d. XIV. Kal. Ian.*, they now appeared, still five days after the Ides, *a. d. XVI. Kal. Ian.* In Macrobius' words: 'with the result that, since the exact date was not commonly known – some observing the addition which Caesar had made to the calendar [thus on the new, but correct date] and others following the old usage [thus two days too late] – the festival came to be regarded as lasting for more days than one'.[60] Macrobius himself goes on to concede that the historical perspective is false insofar that the Saturnalia were already celebrated over several days: but this in itself confirms the authenticity of the reported popular misunderstanding.

As already remarked, the same problem affected private occasions; here, however, the 'sacral conversion', that is to say keeping a constant interval from the Ides, was only an option, not a real obligation. While it was observed by some, others – even among the elite, as demonstrated by the instance of Livia – remained with the old 'date'; Augustus appears to have used this ambivalence to justify a two-day celebration.[61]

[58] Probably L. Claudius (RÜPKE 2008: no. 1160).

[59] Scaevola, *Dig.* 50.16.98.

[60] Macrob. *Sat.* 1.10.2; see also BERGMANN 1984: 13.

[61] SUERBAUM 1980. On Livia 336, on Augustus 334–5 with reference to Suet. *Aug.* 57.1. It is possible – and SUERBAUM's examples suggest this – that actual proximity to the Ides was more likely to suggest such a conversion than a position shortly before the Kalends. Against RADKE 1990 see RÜPKE 1992b and RADKE 1991; cf. FEENEY 2007: 148–60.

The instance of the Saturnalia proves instructive in quite another direction. The celebration over several days cannot hide the fact that, on the calendar, only one day is marked *SAT*, only one day observed by the priests as *feriae Saturni* and *dies natalis* of his temple. Wherever a new edition of the Fasti anni Iuliani was available, there could be no error. This allows us three conclusions: one, the calendar, as a written medium, played no role for the mass of the population; two, the central rituals of the festival gave no impulse worthy of note, or any guiding authority, for general private celebration; three, it is, finally, remarkable that no comparable debate is attested for any other festival. In view of the fragmentary nature of the tradition, such an *argumentum e silentio* has no great value, but it may at least serve heuristically as an indication that only a small number of the *feriae* listed in the calendar had the status of a popular festival, with the consequences shown here.

We must, finally, inquire into the currency of the Julian Calendar beyond Rome itself.[62] In the case of Italy, our inquiry is at least partially satisfied by the examples of *fasti* found throughout the territory, although with a concentration in Latium; Sicily remains problematic.[63] It must, however, not be forgotten that a mass of detailed information about alternative Latin and Italic calendars, either still in use or only recently fallen into disuse, was available to Late Republican and Early Imperial-Period antiquarians;[64] both the Iguvine Tables and the Agram mummy wrappings, which attest to independent Umbrian and Etruscan calendars, may also be no earlier than the century of the Julian reform. The improved ease of use of the Julian Calendar now for the first time made the method of intercalation practised in the city of Rome attractive for the needs of localities beyond Rome; combined with the civil wars and associated social upheavals, and the accelerated integration of the Italian elite instigated by Caesar and Octavian, this factor may have had a positive effect on the calendar's spread. On the other hand, centrally directed exportation of the calendar is nowhere discernible.

Beyond Italy, the calendar of the city of Rome continued to be definitive for central administrative purposes and for the military: the Feriale Duranum proves this, at least for the beginning of the third century CE.[65] In this role, it acquired the function of a standard calendar, called upon to synchronize the various calendars that continued in existence; comparative tables, known as *hemerologia*, existed to serve such purposes of conversion.[66] If the standard calendar ensured that the lengths of local months and years were adjusted to Roman values (often preserving ancient names and/or counting practices), thus making it possible to achieve a consistent equivalence over the years, as in the case of the calendar for

[62] See Rüpke 2007b; cf. FEENEY 2007: 209–11.

[63] For the Fasti Tauromenitani see RÜPKE 1995a: 133–8.

[64] SCHOLZ 1990.

[65] Despite rightly pointing to the hypothetical character of the underlying arguments, REEVES (2004) fails to justify her own claim of a civic festival list with a local bias.

[66] KUBITSCHEK 1915.

the province of Asia, introduced in 9 BCE, in the Greek East the outcome was a very high degree of regulation.[67] In general, the pre-Julian patchwork of calendars actually remained in existence.[68] In the Latin West, on the basis of dated inscriptions and the list of dates from Numidian Theveste,[69] we can assume that the Julian calendar of the city of Rome had a wide distribution. Both the Gaulish calendar from Coligny[70] and the Imperial-Period conversion from a solar to a lunisolar Jewish calendar[71] demonstrate *ex negativo*, by dissociation, the oppressive dominance of the Julian system.

8.2 The Calendar as Collective Memory

It remains the case that we know of only one *fasti* specimen that can be identified with certainty as Republican: the example from Antium. This hardly constitutes a meaningful body of evidence. With the Julian reform, older specimens had become obsolete; in many instances, owners may have destroyed wall-paintings and papyrus-roll calendars, anticipating the ravages of time. In order to move beyond the outcomes of our analysis of Fulvius' *fasti*, therefore, and to be able to define more precisely the social setting of the *fasti* in the Late Republic, we remain dependent on literary testimony. Of this there is little; but, of the few sources available to us, two passages from Cicero assign the *fasti* a central role in Rome's collective memory.

Among the attacks Cicero makes in his second *Philippica* against M. Antonius, he also speaks about the offer of the diadem to Caesar during the Lupercalia of 44 BCE, and stresses what a scandal it is that he, Antonius, who put the diadem on Caesar's head, still lives, while the death of the man who set it aside again meets with general approval. He continues:

> And, moreover, he [*scil.* Caesar] caused it to be recorded in the annals (*in fastis*), under the head of Lupercalia (*ad Lupercalia*), 'That Marcus Antonius, the consul, by command of the people, had offered the kingdom to Caius Caesar, perpetual dictator; and that Caesar had refused to accept it.'[72]

What *fasti*, or parts of *fasti*, may these have been? The inscriptions provide no answer; neither among the historical notes in the consul lists nor in extant calendar

[67] LAFFI 1967; DESSAU 1900; an example from Bithynia: REA 1992; for the Severan Period: EHRHARDT 1984.

[68] See e.g. BISCHOFF 1884, 1919; SAMUEL 1972: 174–87; GRZYBEK 1990.

[69] HERZ 1975 and *CIL* 8.1859.

[70] MONARD 1999.

[71] STERN 2001.

[72] Cic. *Phil.* 2.87. See also Vell. Pat. 2.56.4; Dio Cass. 44.11; App. *B Civ.* 2.109. HOHL 1942; WEINSTOCK 1971: 331–40; GESCHE 1976: 158–61; FRASCHETTI 1985; JEHNE 1987: 316–18; cf. MEYER 1922: 527–8. Translation of Cicero: C. D. Yonge, ed., from the Perseus Project.

entries for 15 February is a note to this effect to be found. Dio, in his parallel account, speaks of *hypomnemata*, by which he perhaps means the *acta publica*.[73] The matter is settled in a passage in a letter by Cicero to M. Brutus, speaking of high honours proposed for D. Iunius Brutus Albinus, who is offering resistance to Antonius:

> But, I know not how, men are more commonly found to be liberal under the influence of fear than grateful in the hour of victory. For I myself, when Decimus Brutus had been delivered, when that day, most joyful to the city, had shed its light upon it, and that very day, as it happened, was the birthday of Brutus, proposed a vote that the name of Brutus should be attached to that day in the calendar (*in fastis ad eum diem*). And in this proposition I followed the precedent of our ancestors, who paid this compliment to Larentia, a woman at whose altar in the Velabrum you pontiffs are in the habit of offering sacrifice. When I proposed this honour to Brutus, my object was that there should be in the calendar a memorial of his most welcome victory.[74]

The reference to Acca Larentia, to whom the Larentalia of 23 December were supposed to be dedicated, leaves no doubt that Cicero's purpose here is an entry in the calendar; this is confirmed for the first passage now by the equivalence between *ad diem* and *ad Lupercalia*. Even here, though, it is worth reminding ourselves that there is no corresponding note or festival of the 'Brutalia' in the extant *fasti*: in both these instances, the intention appears to have remained unfulfilled.

Whereas notices of games (*ludi*) and dedication days had regularly been recorded in the calendar,[75] personal names remained excluded from this portion of the *fasti*. Although the reform did not in itself constitute an initiative actually to change this state of affairs, it does appear to have distinctly lowered the threshold for such changes. A resolution of the Senate in 45 BCE raised the anniversaries of Caesar's most important victories to the status of *feriae*.[76] In contrast to what, earlier, may have been longwinded and controversial discussions about the status of a day such as that of the Saturnalia, the directness of the Senate's decision on this occasion, and its political significance, may have led to a much accelerated reaction on the part of the calendar-makers; the formulations we see in Imperial-

[73] Dio Cass. 44.11.3.

[74] Cic. *Ad Brut.* 23.8 (Translated and edited by J. S. Watson, London: H. G. Bohn, 1862). Cf. Cic. *Phil.* 11.24.

[75] Examples are the *ludi Victoriae Sullanae* (26 October–1 November), and probably also the *ludi Veneris Gentricis*, celebrated as early as 45 BCE (see Cic. *Att.* 13.44.1) between 20 and 30 Quintilis as *ludi Victoriae Caesaris* (they were originally held on the dedication day, 26 September; see DEGRASSI 1963: 485–6; WEINSTOCK 1971: 156; JEHNE 1987: 204–5). Too much has been read into the 'displacement' of these latter games: the relationship of the earlier games on the occasion of the temple dedication to the annual celebration in later years lies in identity of theological orientation, not continuity of an identical cult.

[76] App. *B Civ.* 2.442; Dio Cass. 43.44.6.

Period specimens were very shortly included in these *fasti*.[77] It happened due to force of circumstance, not by command. The only person affected by such a decision, apart from the cult personnel involved in new rituals, was the *rex sacrorum*, whose job it was to announce the new *feriae* on the relevant Nones: this was the official channel for such announcements, and it made a state *feriale*, or 'master calendar', superfluous.

One further point must be considered in respect of the climate of discourse to which both Cicero passages should be assigned. In the year 44, the month of Quintilis, as the month of Caesar's birth, was renamed *Iulius* at the request of the *consul ordinarius* M. Antonius.[78] Determined upon before Caesar's death,[79] the honour was carried through in the context of the cult acts related to Caesar in 'July'.[80] The novelty of the conception is thrown into stark relief by antiquarian discussion, extending from Fulvius Nobilior to Varro, of the association of the names of the months with gods and cult acts. In the context of the contemporary currency of the theme of deification, the idea may well not have been unique; but the reference to a living person, and the background of political conflict, does not permit us to understand the matter solely at the level of theological categories (that Caesar may already have been treated as a god). We are thus led to seek archetypes for this form of distinction in similar Hellenistic practices.[81]

While such renamings of months were subsequently to become part of the standard repertoire of Senatorial modes of adulation, in respect of the veneration of Caesar we may cite a possible directly contemporary imitation. The Cicero letter suggests that the imputed reference of the month *Iunius* to the liberator from monarchic rule actually arose in this context, with reference to the first consul M. Iunius Brutus, although it is not attested until much later.[82] The parallel is all the clearer when we consider that, both in the choice of the month to which Caesar was to lend his name, and in the determination of Brutus' single feast day, the birthday served as central criterion. Caesar's birthday itself – 13 July – was celebrated with state sacrifices: similarly without precedent in Rome.[83]

Two (associated) points distinguish Caesar's instruction with regard to the Lupercalia and Cicero's proposal from *feriae* notices objectively similar in form in extant calendars. Firstly, there is no evidence that either instance ever reached the calendar, and, secondly, neither was based on a Senatorial resolution. We might

[77] 27 March, *Inscr. It.* 13.2.169 (Fasti Verulani): *NP. Feriae, quod eo die C. [Caesar] Alexandriam recepit.* On 6 April *Inscr. It.* 13.2.127 (Fasti Praenestini): *NP. Ludi. F(eriae), q(uod) e(o) d(ie) C. Caesar C. f(ilius) in Africa regem [Iubam devicit].* On 2 August *Inscr. It.* 13.2.191 (Fasti Amiterni): *NP. Feriae, quod eo die C. Caes(ar) C. f(ilius) in Hispan(ia) citer(iore) et quod in Ponto eod(em) die regem Pharnacem devicit.* And *ibid.* on 9 August: *NP. Soli Indigiti in Colle Quirinale. Fer(iae), q(uod) e(o) die C. Caes(ar) C. f(ilius) Pharsali devicit.*

[78] Suet. *Iul.* 76.1; Censorinus, *DN* 22.16; Macrob. *Sat.* 1.12.34; Flor. *Epit.* 2.13.91; Talbert 1984: 360–2.

[79] Dio 44.6.2; App. *B Civ.* 2.443.

[80] Dio Cass. 45.7.2; App. *B Civ.* 5.404.

[81] Scott 1931; Weinstock 1971: 152–8.

[82] Macrob. *Sat.* 1.12.31. Generally assumed in later Neronian times: Tac. *Ann.* 16.12.2.

[83] Weinstock 1971: 206–12; Dio Cass. 44.4.4. Cf. Dio Cass. 51.19.2; *CIL* 12.4333 *(lex arae Augusti Narbonensis)*.

almost suspect that they exemplify the excess born of a political climate in which, on the one hand, ideas of novel ways to honour individuals proliferated to an inflationary degree, while, on the other, movement appears to have been brought to the calendar by a great reform that was still fresh. Amendment of the calendar, this venerable document up to now regarded as immune to all change, marked the frontier of the thinkable. But even the imagination remained subject to limits. Only events of supposedly secular consequence had a place in the calendar: the removal of the monarchy, and the refusal to reintroduce it.

The proposal to establish 'Brutalia' is not the first occasion on which Cicero speaks of the *fasti*, or of changes to the *fasti*. Disregarding the passages concerned with Cn. Flavius or with contemporary manipulations of terms of office, however, those passages written prior to the Julian reform share a common characteristic: by *fasti*, Cicero always means the consular lists;[84] they represent the tangible form of the collective memory, a scroll of honour. Interestingly, this 'pre-Julian' premise corresponds to the literary message of a later age. Down to Claudius Claudianus' manifold use of the motif, the theme is always the same: the personal honour signified by the presence of one's name's on this list.

8.3 Augustus and the Power of Dates

We need to correct the picture of the social presence of the *fasti* that has emerged from our analysis of the literary tradition. Historically speaking, the practice reflected in the Cicero passages of 44/43 actually inaugurated a period of heightened intensity so far as work on the calendar is concerned. To the end of Augustus' rule alone, Senate resolutions created some thirty new *feriae* in rapid succession, thus virtually doubling the number of 'holidays'. So far as we can tell from the ancient material that remains, this marks the beginning of the most productive phase in calendar production. The period of increased activity was accompanied by the production of antiquarian commentaries. These include late works of Varro and the treatise by Cincius. The *in-situ* commentary on an actual calendar specimen (the Praenestini), and Ovid's verse commentary,[85] may be understood as attempts to reach beyond the narrow readership enjoyed by antiquarian works. Iulius Modestus and Masurius Sabinus continued the genre in the Tiberian period, in a line that extends via Macrobius and Iohannes Lydus to the end of the Western Empire and beyond.[86]

Here too, however, a reservation must be entered. The calendar's triumphal progress under Augustus did not continue uninterrupted. Even though a few

[84] Cic. *Sest.* 33; *Pis.* 30; *Fam.* 5.12.5.
[85] RÜPKE 1994.
[86] DEGRASSI 1963: xxv–vi.

recently discovered inscriptions attest to the genre's survival beyond the first period of the Principate, and calendars in painted and in written, book form still adorned Roman houses and 'desktops' in the third and fourth centuries, it cannot be denied that a definite collapse occurred as early as the post-Tiberian period. While there is a demonstrable, continuous rise in the production of other epigraphic genres into the Severan period, the number of calendars – less so of magistrate lists – declines rapidly. The reasons for this decline must be investigated. Initially, however, we need to pursue further and define more precisely the results gained from our initial survey of the various extant calendars, and, having regard for the dominance of the consul lists in the panegyric literary tradition, more closely determine the reasons behind the rise in the genre's fortunes.

Everything that, in the Late Republican period, helped the consular list to the popular standing indicated here, and discernible in rhetorical reflection in Cicero, was put into question by the political upheaval instigated by Caesar. The consulate itself gradually underwent re-evaluation, as its occupancy no longer reflected real power relationships; rather than the office, the object of competition was now the monarch's favour, and it was only at a further remove that this favour found expression in the occupancy of the highest magistracies. Although the consulate continued to be an office of the highest prestige, the establishment of regular and multiple suffect consulates, together with the development of an autonomous Imperial administration, made the mere gift of the consulate appear no longer adequate as the highest official status an individual could achieve. Assessment of one's social standing, not to speak of one's own interest in adequate self-representation, demanded more diverse documentation of a career, to include initial offices held, priestly and post-consular offices, and perhaps even the low age at which such offices were attained (*suo anno* in Republican parlance). Only the fiction of the continuity of the Republican consulate assured the continued existence of the consul lists.

In a certain sense, it was that very fiction of the Republican consulate that assured the lists any kind of continued existence at all. Upon examination, Cicero's early testimonies in this regard are already negative, lamenting as they do that the objective distinction accorded the office-holder is inappropriately deficient, having regard for his actual performance. The idea of a consulate filled from the ranks of the Senate, and occupied with its consent for a year by a member of the nobility who then withdrew again into the body of the Senate, was inappropriate by the second half of the second century BCE.[87] The term of office increasingly became a route to other distinctions, calculated precisely to avoid merely returning to the bosom of the older consularies. For the individual, the mere unadorned fact of being recorded in the general list accordingly tended to lose its character as an ultimate honour, and, during the following Augustan period, when honorific inscriptions were erected at an unprecedented rate, the list as such ceased to serve

[87] See HACKL 1982. Cicero: *Sest.* 33; *Pis.* 30; *Fam.* 5.12.5.

that end. In contrast, the calendar's honorific component, the *dies natales*, appeared in a more favourable light. Now as before, temples and other building projects symbolized their founders' success in war. Escalating ambitions were no longer satisfied by *aedes* and *aediculae*, however: major buildings and entire forums were the order of the day. The egalitarianism of calendar entries, and the indirectness of the reference to the person behind the foundation, and to the achievements that, alone, legitimized temple-building, may have led to a perception of the marble *fasti* as an excessively ponderous medium.

This perception was, however, overcome where the large number of new structures brought real movement to the calendar, and the account in Augustus' *Res gestae* provides the clearest proof that such a situation did in fact emerge. His achievements in this regard are listed not only in the main text, but also in the final summary: temples of Mars, Jupiter Tonans, Jupiter Feretrius, Apollo, Divus Iulius, Quirinus, Minerva, Juno Regina, Jupiter Libertas, the Lares, the Di Penates, Iuventas, a *pulvinar* of Mater Magna near the Circus maximus, and further structures, including eighty-six restorations, the Capitol among them.[88] Not only the new structures, but also many restorations, were associated with new or altered dedication dates: within a short period, a distinct change ensued to the picture of the year as presented in the calendar, reinforced still further by activities instigated by Octavian on the part of his supporters.[89]

Despite their great number, the Augustan temples did not simply represent elements of a major architectural programme. As in the Republican Period, many of the cult structures, and these the most important ones, resulted from specific historical events, often military successes, to which they referred in their decor. Thus, in recording these events, the calendar again acquired a historical dimension, even though in a largely contemporary context. In this respect, it was never more topical than in the Late Augustan Period, and surely served as a stimulus to Ovid.

This reflection of contemporary events was reinforced by the many new *feriae*, which give Imperial-Period calendars their characteristic look. They represented a sea-change, not only to the Fulvian *fasti*, but also to the Flavian tradition, and changed both the calendar and the ritual face of religion in the city of Rome. At first glance, it seems that the Republican conception of the holiday as assigned to a divinity, as in *feriae Iovi*, was relinquished, to be replaced by a way of thinking that permitted *feriae* to be established with no theological reference, in honour of (living) individuals.[90] It would be astonishing if so radical a break in a central religious institution had occurred without opposition, and leaving no traces. In fact, such traces do exist.

The normal formulation for an Imperial feast day in the *fasti* is *Feriae ex senatus consulto, quod eo die ...* Abbreviated, this reads *FER EX SC QED.* DEGRASSI integrates

[88] August. *Gest.* 19–21; *summa*. For the concept of 'restoration' see GROS, SAURON 1988: 60–1 and 176 n. 33.

[89] Suet. *Aug.* 29.4; GROS 1976: 32–3: 21 confirmed new or altered dates.

[90] FRASCHETTI 1988: 121–33.

this wording in his calendar edition wherever there appears to be adequate space. Concentrating our attention on the extant portions of calendars, we can discern at least two calendars that use a shorter wording as well as this one. In the Fasti Amiterni, against seven certain instances of the long version, four offer a simple *feriae quod eo die* (the abbreviations may be disregarded in the following). In the Fasti Verulani, the ratio is three to one.

Are these, as the similarity of the formulations suggests, and, to my knowledge, has always been accepted, merely meaningless variations between a short and a long version, dictated by the space available? A glance at the festivals thus presented reveals that this is by no means the case. The *ex senatus consulto*, the reference to the Senate's decision, is absent from the Fasti Amiterni for Caesar's *dies natalis*, celebrated on 12 July, from Caesar's victories in Spain in 49 and over Pharnaces in 47, both celebrated on 2 August, and from his victory at Pharsalus on 9 August;[91] the Fasti Verulani list Caesar's capture of Alexandria on 27 March with the short version:[92] so, are we dealing solely with dates concerning Caesar here? In the calendars cited, this is contradicted by the entry in the Fasti Amiterni for 10 August, celebrating as *feriae* from 7 BCE the dedication of an *ara* to Ceres and Mars.[93] Coming as it does directly after the Caesarean date of 9 August, this could conceivably be an error. The wording in the Fasti Antiates ministrorum fails to follow the customary pattern for these festival notices, and their entry for the same day in any case suggests an exceptional case from another context: *Feriae Cereri et Opi Aug(ustae)*[94] belongs with this calendar's explanation of the Republican *feriae*, and does not directly concern Imperial festivals.

The conclusion is that *ex senatus consulto* is omitted only in the case of dates that concern Caesar, and that were made *feriae* between 45 and 42, the last occasion being the celebration of his *dies natalis*. No counter-arguments arise from other *fasti*, if we disregard false integrations in the editions.[95] The total number of instances is sufficient to assume an established rule, whose origin and intention must, however, remain obscure.

Further help is available in this situation from what must be regarded as the oldest Augustan calendar, the Fasti fratrum Arvalium. They have the following entries for Imperial festivals:

> *F(eriae) ex s(enatus) c(onsulto), [q(uod) e(o) d(ie) Imp. Caesar rem pu]blic(am) tristiss(imo) p[e]riculo [libera]vit* (1 August, *feriae* since 30 BCE).[96]

[91] *Inscr. It.* 13.2.189–91.

[92] *Inscr. It.* 13.2.169.

[93] *Inscr. It.* 13.2.191.

[94] *Inscr. It.* 13.2.208.

[95] E.g. *Inscr. It.* 13.2.147 (Fasti Vallenses 12. 7.).

[96] *Inscr. It.* 13.2.31; there too the following entries (31–5). For the dates of origin see DEGRASSI's '*Commentarii diurni*' (1963: 388–546).

[Feriae[97] *C. Caesa]ris h(onoris) c(ausa) Hisp(ania) [citerior]e devicta [et quod in P]onto regem [Pharnace]m dev[i]cit* (2 August, *feriae* since 45 BCE).

Feriae Imp. Caesaris h(onoris) c(ausa), quod eo die vicit Actium; four letters *ex s(enatus) c(onsulto)* are added as a correction over the beginning of *Imp.* (2 September, *feriae* probably since 30 BCE). Dio reports that the *feriae* decision had been arrived at on the day the news of the victory arrived; this would be a further analogy with the institution of the *supplicatio.*[98]

Feriae et supplicationes ad omnia pulvinaria, q(uod) e(o) d(ie) Caesar August(us) in Sicilia vicit (3 September, *feriae* since 36 BCE).

F(eriae) ex s(enatus) c(onsulto), q(uod) e(o) d(ie) Imp. Caesar Aug(ustus) pont(ifex) ma[x(imus)] natus est (23 September, *feriae* since 30 BCE).

The evidence of the Fasti fratrum Arvalium both confirms and adds precision to that of the later calendar. The formula '*feriae* by decision of the Senate' is absent in the case of Caesar's memorial days, and, under or perhaps at the instigation of his successor Caesar Divi filius, there occurs a terminological sea change, leading, via an intermediary stage in 36 BCE, to the formula, canonic by the end of the thirties, emphasizing legitimization by the 'Republican' institution of the Senate.

The fact that the calendar preserves the older wording for the earlier festivals through into the late Tiberian period – with the terminological disintegration discernible in the Fasti Antiates ministrorum – shows the *fasti* attempting to reproduce the wording of the Senate decisions or decrees that led to the establishment of the *feriae* in question. This underlines the conclusion expressed earlier, that there was no central-Roman edition of the *fasti* concerned with safeguarding religious interests: the *fasti* merely attempted to reproduce norms formulated elsewhere, in the pontifical college or the Senate. The concern for terminological consistency, in what appears to be an inconsequential detail, and in contradistinction to the practice in many other areas, shows that this particular aspect was regarded as important, and was to a certain extent subject to social control.

If the above considerations are true, it is all the more remarkable that one terminological detail present in the Fasti fratrum Arvalium is missing in later *fasti*. This concerns the expressions *feriae C(ai)* and *Imp(eratoris) Caesaris h(onoris) c(ausa)*, used respectively on 2 August for Caesar, and on 2 September for Octavian. The wording does not apply to later Imperial festivals in the same calendar, and so cannot be an idiosyncrasy of this particular edition; moreover, for the celebration of the victory at Actium it is corrected above the line by Octavian's Senate formula. The expression *honoris causa,* clearly also suppressed for the Caesarean *feriae* in later *fasti*, enables us to gain a deeper insight into the theological dimension of

97 DEGRASSI integrates here: *ex s(enatus) c(onsulto).*

98 See DEGRASSI 1963: 505. Dio Cass. 51.19.1–2.

the Imperial feasts, discussed earlier, and allows us to reconstruct its evolution in more detail.

The *feriae* defined in Caesar's lifetime and for a short time afterwards (42 BCE) – all of them on the basis of decisions of the Senate, as is shown by the tradition in Dio – were *feriae C. Caesaris honoris causa*. These were not *feriae Caesari*, analogous to the *feriae* of the gods (e.g. *feriae Iovi*), but the terminology points strongly in that direction. This is demonstrated by the terminological treatment of *supplicationes*, held in token of gratitude after victories.[99] Such gratitude is directed to the gods (*ut dies immortalibus habeatur honos*) 'in the name of the commander' (*nomine imperatoris*), but, in its duration of one, two, three, eventually even twenty or fifty days, the festival of thanks directly honoured the commander.[100] In this context, therefore, notwithstanding the underlying tendency of the ritual, in the forties of the first century BCE *honos* still refers to the theological principle, the divine addressee.

Neither can this connotation of the expression *alicuius honoris causa* in the context of a cult celebration be ignored in respect of the *feriae*. A form of words was found that specified Caesar as the addressee of undoubtedly godlike honours, while circumventing the full deification implied by the dative locution *feriae alicui*. The *feriae* resolved upon for him thus augmented an entire arsenal of similar religious honours.[101] The fact that the naming of Caesar's *dies natalis* as *feriae* in 42 BCE[102] did not address *divus Iulius*, but, once again – by the testimony of the Fasti Amiterni – *C. Caesar*, appears to me to be of central importance in the evolution of this process:[103] no retreat had been made from the theological impetus of 45/44; on the contrary, the direction embarked upon there had been confirmed, and was available to be followed further.

After the victory won over Sex. Pompeius for Caesar's heir, new measures were taken in 36 to honour him by the declaration of special *feriae*.[104] However, either the Senate or Octavian shrank from directly adopting the Caesarean model. The theological conceit now chosen – *feriae et supplicationes ad omnia pulvinaria, quod eo die Caesar ...*[105] – demonstrates the conceptual link to the *supplicationes* already mooted here for Caesar. Now, however, by the link with *supplicationes ad omnia pulvinaria*, the *feriae* were given an unambiguously 'correct', that is to say supernatural, theological reference. This was, however, expressed as non-specifically as in the case of the thanksgiving celebrations, addressed to all 'immortal gods'. Against this indeterminate background, the actual, personal reference of the occasion for celebration was brought all the more clearly to the fore.

99 For a general treatment see HALKIN 1953; RÜPKE 1990a: 215–17; NAIDEN 2006.

100 RÜPKE 1994c, 1990a: 216.

101 WEINSTOCK 1971; cf. SCHEID 1993.

102 Dio Cass 47.18.5–6.

103 *Inscr. It.* 13.2.189.

104 For subsequent celebrations of this year (e.g. *domus publica*) see FRASCHETTI 1990: 91–3.

105 See *Inscr. It.* 13.2.33 (Fasti fratrum Arvalium 3. 9.).

Although the formula for 3 September is maintained in later *fasti* (those from Amiternum in particular), the experimental form subsequently ceases to be further pursued. The formula used for the celebration of the victory at Actium, in 30 BCE, is the one that then became established as the norm. The theological reference fell entirely out of use, the resulting gap being filled by the entirely different expression *ex senatus consulto*: political legitimization took the place of theological clarity, and the problematic theology of *feriae* of all the gods, conjured by the formula used in 36, was set aside.

It is for these very *feriae* of 2 September that we find in the *fasti* in the Arvals' sanctuary a peculiar hybrid formula, where an *ex senatus consulto* has subsequently been added to the *honoris causa*. Considering not only the doubts that had led to the search for another solution as early as 36, but also the 'normal' form provided on 1 August for the *feriae* of 30 BCE, there is probably no need to interpret this as intimating that Augustus or like-minded individuals at first decided for the old *honos* formula, and so to assume that the correction was added later, perhaps in 27 BCE. At the time the calendar was created, Caesarean entries were still in the clear majority,[106] so it is not surprising that the stonemason spontaneously adopted the *honoris causa* formula; but, as a grave material error (and a serious embarrassment), this had to be corrected: the festivals associated with Octavian / Augustus were – in official parlance – held not 'in his honour', but merely 'because he accomplished this or that on this day'.[107]

Notwithstanding the narrative outlined here, it was not unheard of for new 'theological' *feriae*, or *feriae alicui deo*, to be created in isolated instances, at least one of which can be proven. The death of Agrippina fell on 18 October, the day on which Seianus had been executed two years before (31 CE). By no means, however, was the day now declared *dies vitiosus*: this classification represented a unique occurrence in the early Augustan period; despite the Senatorial formula (*vitiosus ex senatus consulto*), it was not repeated. The day was also not simply declared *dies nefastus*[108]; the loose employment of this term elsewhere[109] has nothing to do with the vocabulary of sacral law and the strict terminology of the calendar. What Tiberius did with this day emerges unambiguously from historiographical and calendrical sources. Tacitus reports that it was decided 'to dedicate 18 October, day of death twice over, to Jupiter for all time', and the Fasti Filocali note this day with the addition *Iovi Liberatori*.[110] Having regard for the notice in the *fasti*, Tacitus' wording – *donum Iovi sacrare* – must be seen as a form of words for the establishment of *feriae Iovi Liberatori*: and at the same time as testimony to the survival of the Republican conception of *feriae*.

106 DEGRASSI 1963: 369.
107 Against FRASCHETTI 1988, *passim*.
108 Thus HERZ 1978: 1158.
109 Tac. *Ann.* 14.12.1.
110 Tac. *Ann.* 6.25.3; cf. Dio Cass. 58.12.4.

At least so far as the city of Rome is concerned, the cult repertoire of the new festivals (like those of the Imperial cult in general) remained within the bounds of custom, allowing the habitual scope for modifications. Priesthoods performed sacrifices (*immolationes*) and supplicatory processions (*supplicationes*); vows (*vota*) were made, and redeemed by the slaughter and consumption of appropriate sacrificial beasts (*hostiae*).[111] Orientation to the Imperial House, the *domus Augusta*, might be acknowledged quite straightforwardly, by the addition of locutions such as *pro salute imperatoris* to prayers, or by endowing the divinity addressed with the epithet *Augusta* (or *Augustus*).[112]

Dates connected with Augustus included, besides his birthday, memorial days for great victories, anniversaries of dedications (which might themselves be connected with victories), celebrations of his principate, inaugural days (as first consul, supreme pontiff, *pater patriae*), and, finally, important events from his life and from the lives of members of the Imperial family.[113] Not all days had the character of *feriae*: many dates are known to us only from texts other than calendars. If January provides more Augustan dates than any other month, this is due to the fact that the Fasti Praenestini have survived complete for this month, for which their commentary section offers many entries that were probably never to be found in calendars.

Although it misrepresents the calendrical picture, the evidence of the dates from Praeneste is nevertheless highly informative. On the one hand, it demonstrates the density of the structure that had evolved, and on the other the gradations within the long list. Not all dates are declared *feriae*, and these do not all become popular festivals. The 'popular' character of a festival is nearly always associated with the presence of *ludi*. This condition is met only on Augustus' birthday with games on 23 September,[114] the Augustalia preceding the dedication day of the Ara Fortunae Reducis (12 October),[115] and the *ludi Martiales*. The latter were celebrated up to Filocalus' time on 12 May,[116] but, for a while at least, also on 1 August, the main Augustan Victory Day;[117] both dates seem to relate to the *constitutio* and *dedicatio* of the temple of Mars Ultor. To these must be added the Arvals' moveable *(conceptivi) ludi* during the multi-day festival of Dea Dia.[118]

That these *feriae* are not distributed as evenly as possible across the calendar, notwithstanding the sheer weight of numbers of the days involved, arises from

[111] See WISSOWA 1912: 409–32.

[112] SCHEID 1990: 285–676.

[113] HERZ 1978: 1148 f.; EHRENBERG, JONES 1976: 44–55.

[114] *Inscr. It.* 13.2.80 (Fasti Maffeiani 23. 9.).

[115] The Fasti Amiterni call them *ludi divo Augusto et Fort(unae) Reduci*, and have them begin on 5 October (*Inscr. It.* 13.2.195), while in the Fasti Antiates ministrorum, which are probably not much later, they begin on 3 October (209).

[116] *Inscr. It.* 13.2.76 (Fasti Maffeiani). 247 (Fasti Filocali).

[117] DEGRASSI 1963: 456; WEINSTOCK 1971: 157; HERZ 1978: 1149.

[118] SCHEID 1990: 636–9.

the strategies underlying their placement. Many temple dedications occur on 23 September, Augustus' birthday,[119] while others, along with triumphs, take place in August, 'his' month. Livia's birthday on 30 January was indirectly made a state holiday – *feriae* according to the Fasti Praenestini[120] – by the dedication of the Ara Pacis Augustae in 9 BCE.[121]

Giving additional meanings to particular days, and, conversely, raising the profile of events by appropriate timing, was nothing new. As a principle, it is already found in the structure of the Hellenistic ruler cult,[122] while Caesar provided a model for Rome by his machinations during the Civil War. It appears that he did not leave the arrival of his messengers in Rome to the vagaries of chance, when they brought news of the victory at Munda on 15 February 45: they arrived with the tidings that brought the Civil War to an end on the eve of the Parilia, immediately before the 'birthday of Rome', 'giving Caesar's victory the aura of a new foundation of the city, and himself the character of its mythic first king Romulus'.[123] Characteristic of the continued exercise of these practices at later periods is that they no longer attached to the old festivals, but to Augustan dates, or to dynastic anniversaries underlying those dates.[124]

In our analysis of the temple dedications of Fulvius and Aemilius, reference was made to strategies involved in the placement of dates, focusing on the necessity of reconciling the degree of publicity obtained with the problem of divided attention. Neither the configuration nor the placement of Augustan memorial days reveals any consistent concern to procure an optimum degree of public participation in respect of all such days.[125] Even if the original celebrations and dedications were carefully and thoroughly planned (an impression may be gained from the frieze of the Ara Pacis Augustae, although it fails to provide a protocol for the procession), accounts of repeat celebrations are thin on the ground. This is not to say, at least until the death of Augustus, that there was no cult commemoration; at the same time, it provides no encouragement for the idea of lavish annual celebrations. Over and above the original celebration, and routine ritual, which may very well, in less important instances, have been maintained by professional personnel such as temple wardens (*aeditui*) or slaves (*publici*), the establishment of an annual memorial day would take on a symbolic function, which might be thought of as not dependent on participation in the ritual, and on buildings. The vehicle that might then come most easily to mind as an alternative means of publication, made necessary by the absence of a public ritual element, is the calendar.

119 Fasti Palatii Urbinatis (*Inscr. It.* 13.2.63); GROS 1976: 33.
120 *Inscr. It.* 13.2.117.
121 HERZ 1978: 1152; ALFÖLDI 1966; BOSWORTH 1982.
122 HERZ 1978: 1173–4; GZRYBEK 1990; CHANIOTIS 1991: 136.
123 JEHNE 1987: 204; cf. Suet. *Cal.* 16.4.
124 HERZ 1978: 1156–7, 1165, 1172, 1176.
125 Cf. HERZ 1978; Suet. *Cal.* 16.4.

Two further elements merit consideration here. The first concerns month-names. The aura of tradition surrounding the names of the months would add particular weight to a step such as the one just outlined. This is entirely consistent with the renaming of the months *Quintilis* and *Sextilis* as *Iulius* and *Augustus*: on the one hand, the old, numeric names had less intrinsic substance; on the other, the exceptional character of the situations that underlay the changes, and at the same time made them enforceable, was in this way clarified.[126]

Attempts subsequently to include such name-changes in the canon of honorific measures normal for new emperors proved unsuccessful. Tiberius rejected the proposal to rename September and October after him and his mother.[127] When Caligula renamed September *Germanicus* in honour of his father of that name,[128] it led to no discernible amendments in the calendar. Other attempts that proved impossible to implement were those by Nero, which extended even to the family of 'divine' months,[129] and Domitian's renaming of September and October as *Germanicus* – this time in reflection of his own recently assumed victory name after the campaign against the Chatti – and *Domitianus*, as well as many more whose realization was never pursued.[130]

Having regard for the originators of such plans, their presumption and thus inevitable and unmarked end, and posterity's opinion of them, the problem here might appear to be one of acceptance. Similar considerations apply to adulatory schemes for changing the time of New Year.[131] The central point here is not the question of failure or success, but the paths by which such name-changes could be realized at all, and how their propagandist effect might unfold. Their implementation required neither great expenditure of ritual nor built structures, but merely the use of the name of the month in dates; this could happen nowhere more effectively than in a document that provided the norm for all dates: the calendar.

The second point confirms the relatively low ritual status of most of the memorial days. The same period that produced them also introduced into Rome a practice common in the Hellenistic world: festivals celebrated at intervals of more than one year. The earliest testimonies reappear in the traditions in connection with the exceptional honours awarded Caesar during the final year of his life. While Appian speaks of five-yearly vows, to be offered him publicly by priests and priestesses (presumably the *virgines Vestales*), Dio Cassius mentions games at five-yearly intervals.[132] Both probably relate to the same institution: the one from the

126 Petron. 53.2 still uses *Sextilis* (K. F. C. Rose 1971: 27–8).

127 Suet. *Tib.* 26.

128 Suet. *Cal.* 15.2.

129 Tac. *Ann.* 16.12.2.

130 Macrob. *Sat.* 1.12.36–7. *Domitianus*: *Lex Irnitana* 10 C 40; *Germanicus*: Mart. 9.1.4; see also Plut. *Numa* 18.4–19.4; Dio Cass. 73.15.3; Scott 1931; Buren 1923: 163: *VII Kal. Commod.*; *CIL* 14.2113 = *ILS* 5193.

131 Tac. *Ann.* 13.10.1; but cf. Suet. *Aug.* 59.

132 App. *B Civ.* 2.442; Dio Cass. 44.6.2.

standpoint of the *votum*, the other from that of the *voti solutio* due five years later. Roman archetypes may be discerned in the *vota quinquennalia* and *decennalia* of the period of crisis during the Second Punic War; these were for the state (*pro re publica*), but did not continue long in practice. Augustan sources give reason to think that more recent models may have existed in the Greek world, already related to the Roman ruler in the person of Caesar himself.[133]

Although the Ides of March forestalled redemption of the vow to Caesar, ritual cycles of this kind were realized at intervals of ten years, and in the times of the short-lived soldier emperors five years.[134] A series of *vota quinquennalia* and *ludi quinquennales* for Octavian, subsequently Augustus, was the subject of a decision in 30 BCE; they were organized in rotation by the consuls and the four highest priestly colleges.[135] These exceptional events, and other, even longer-term cycles (the *ludi saeculares*, which, subject to political requirements, were not always separated by a full *saeculum*),[136] were celebrated with lavish cult activity, and the fullest possible degree of public mobilization.[137]

The Principate's ritual staging was the subject of careful planning, which, despite the great number of events commemorated, appears to have concentrated mainly on the creation of memorable high points. With due regard for the great size of the audience, the action did not lose itself in countless rituals, which would never in any case have reached a regular, broad public. Such constant memorialization was much better provided by everyday moments in the smallest of circles, above all in the cult of the *genius Augusti*, which was integrated into the Lares cult. But we must not allow this aspect to distract us from an intermediate level, a multitude of *collegia* and *sodalitates* and *sacerdotes publici* who commemorated and celebrated those countless events and rituals: a central component of the religion of the Principate, a natural component of the life of the leading elite, and object of its aspirations, both locally and Empire-wide, and an important factor in its cohesion.

8.4 The Calendar as Roman Breviary

The inclusion of those Imperial festivals enjoying a 'foundation', and the changes in the area of the *dies natales templorum*, contributed to a substantial increase in the historicization of the *fasti* first discernible in the work of Fulvius. How did the

133 WEINSTOCK 1971: 310–17.

134 See CHASTAGNOL 1987, 1988a, 1988b; HEIM 1992: 267–306.

135 August. *Gest.* 9; HEINEN 1910: 15–16 with n. 1.

136 PIGHI 1965; SCHNEGG-KÖHLER 2002.

137 See the *commentarii ludorum quintorum*, PIGHI 1965: 111, Z. 53–4: *quos [quod spectare quam plurimos convenit] propter re[ligione]m atque etiam quod tali spectaculo [nemo iterum intererit.*

feriae, which, behind their ancient ciphers, made up the age-old substance of public religion, relate to this development?

Broadly speaking, we know little about the festivals that lay behind the Flavian abbreviations. The *fasti* themselves offer nothing beyond the information that these were *feriae*; not even the names behind the abbreviations provide reliable guides. Only a few dates were of sufficient popular interest for details to be mentioned in letters, speeches, or works of literature. Almost our only sources of information are commentaries or systematic antiquarian literature, of which only fragments survive. Ovid's verse commentary is the most substantial relevant body of text. Unlike the regular *acta* (protocols), however, of which we have inscribed excerpts for the *ludi saeculares* and the Arval brethren,[138] these texts rarely provide detailed descriptions of the rituals involved, but tend rather to concentrate on providing interpretations of the festivals' ritual and theology.

MARY BEARD has pointed out, to wide agreement, that, while it is permissible to use such texts as sources for the reconstruction of ritual, we must also read them as autonomous 'theological' works, which, rather than making statements *about* Roman religion, actually form part of it. Ritual and exegesis must be considered in close association: it is the capacity of ritual to stimulate interpretations that determines its continuing vigour. It is only when thus seen in combination that the ritual appears as 'part of a symbolic, religious discourse that continued to be meaningful in the complex urban society of Rome in the age of Cicero, Augustus, Seneca, or Hadrian,' instead of representing a functionless survival.[139]

If we concentrate on the exegeses of those festivals portrayed in the *fasti*, while keeping in mind the interpretive context just outlined, it becomes clear that historical interpretation plays a substantial role. A series of festivals and associated rituals are represented as institutions going back to Rome's early period. The immediate foundation period, Romulus, and then Numa, dominate.[140] The status of the exegeses, as defined above, permits us to suggest that the annual feasts evoked a particular and distinctly selective conception of Rome's past, and in this sense possessed a historical dimension. But what applies to the festivals themselves would also have to apply to their portrayal in the calendar, and all the more so to the synopsis of dates established there: the *fasti* too, as a function of recording the ancient (and yet topically interpreted) festivals, would have drawn a historical picture, serving as well to legitimize current power relationships as to assure their own identity.[141]

[138] SCHEID 1990b; RÜPKE 1993b; SCHEID 1998.

[139] BEARD 1987: 1; see FEENEY 1998; ELM, RÜPKE AND WALDNER 2006; RÜPKE AND SPICKERMANN 2009.

[140] WALLACE-HADRILL 1988: 226.

[141] BEARD 1987: 1. Cf. 11: 'The calendar was one way (and a changing way) of representing "Roman time".' WALLACE-HADRILL 1988: 226: 'The calendar, by a non-chronological re-enactment of Roman history through the course of the year, represents a sense of Romanness.'

In the process of transference from the exegetic texts to the *fasti*, the interpretation proposed acquires a dimension that has failed to be adequately reflected by those who put it forward. It remains undisputed that the learned exegeses cannot be assessed according to the criterion 'true or false'. A Varro or an Ovid may well make historical or historical-linguistic claims that are, in principle, incapable of being confirmed or negated. This is, however, immaterial, if we regard these writers as participants in a symbolic discourse. Not even a criterion of convergence can usefully be employed to assess the material: there is no 'official' interpretation capable of offering a standard for judging individual formulations.[142] At first glance, not even consensus had a part to play: the question as to whether this or that interpretation was capable of commanding contemporary consensus has no effect on the determination that this specific interpretation belongs to the general spectrum of the 'innate' world view.

All of this applies to the literary output of individual Roman 'theologians'. Including the *fasti* as just one of a multitude of examples of realized visual symbols alters the nature of the problem: what interests us now is no longer the view of the originator, but that of the recipient. To describe this type of calendar as a 'symbolic portrayal of Roman history' can be regarded as a sensible statement only if we assume a sufficient reception (or give grounds for the failure of a relevant intention on the part of an originator).

Certain considerations arouse doubts as to whether we may simply equate the content of expert exegeses with the average Roman's reception of the *fasti*: as our discussion of the festivals of the *domus Augusta* has already shown, there is nothing we can usefully say about the popularity of most festivals, and the audience attracted by them. This audience was certainly fragmented on the basis of sex, social group, and profession. The bakers did not frequent the festival of the Tibicines, and women's festivals were barred to men. A corresponding social fragmentation no doubt also affected information about such festivals: knowledge of rituals, and of their current interpretations, is unlikely to have extended beyond festivals one had oneself attended, or that had been attended by a significant other person. Even so far as those attending are concerned, the question of communication about the ritual must be judged sceptically: did spectators at chariot races discuss Romulus and the Sabines, or the leading horse and the green team's driver?[143] If an intellectual conversation nevertheless developed, the hierarchical seating plan[144] prevented wider oral diffusion of any such ideas.

There is no evidence either that an explanatory address was integrated into the ritual itself. While, in the case of the secular games, the historical reference could easily be, and was in fact, included in the texts of invitations and prayers, there is

142 On the problem of the social control of comparable interpretations and traditions see VANSINA 1985: 94–123.

143 See Ov. *Ars am.* 1.143–6; cf. FRIEDLÄNDER AND WISSOWA 1920: 3–9, 29–40; CLAVEL-LÉVÊQUE 1984, 1986.

144 See KOLENDO 1981; *Lex Irnit.* 9 A 21–8 (with GONZÁLEZ 1986: 226); *Lex Urson.* 125–7.

no indication that the same applied to other occasions. A variety of entirely different historical references was provided even for such massively and regularly interpreted events as the Regifugium and the Poplifugia.[145] This is not to say that there were not cult legends, whose tradition might circulate outside the ritual, or in loose association with it, in the form of dramatic presentations[146] near to the time of the celebration.[147] But the variability of the interpretations in circulation makes a conceptual link with the ritual, and the dominance of any interpretation provided in such a context, all the more precarious.[148]

Alternately, the influence of antiquarian texts must not be overestimated. Compared with the audience enjoyed by *fasti*, the readerships of Varro, Verrius, and even Ovid, can have constituted only a vanishingly small proportion. It cannot be demonstrated that the 'symbolic discourse' engaged in by these writers involved large numbers of individuals. This is not to say that their reflections represented mere idiosyncrasies. From the last years of the Republic onwards, Romulus played a substantial role in political imagery; and reference to the early period, with Augustus as the *alter Romulus*, also found expression in programmes of statuary, with the *forum Augustum* as their public showcase.[149] There is no suggestion here that the same applied to the traditional festivals, outside exegetic literature.

It is always problematic to draw conclusions from this literature. The exegetic pressure to which each individual element of the ritual was exposed by the particular context of the genre treating it, together with the narrow available repertoire of explanatory strategies (etymology, mythic archetype/portrayal (typology), genetic explanation), created meanings that were distorted and greatly inflated in comparison with the spontaneous, contemporaneous interpretations of participants. Notwithstanding what has been said up to now, it remains undisputed that such interpretations as these existed, even if they were not verbalized, or otherwise communicated. Ritual is linked to the culturally conditioned, imaginative world of every individual; the components of the ritual are 'understood', beyond explicit texts and prayers; individual gestures are 'grasped' pragmatically, against the background of the entire apparatus of gestures of greeting, submission, giving, sharing, eating.[150] By its very nature, this level of interpretation leaves no trace, which is why the extreme cases cited are nevertheless important, if we are ever going to have any chance at all of configuring the spectrum of possibilities.

Looking at the outcome of this discussion in the light of the problem initially posed, we find ourselves faced with a largely negative conclusion. These exegeses

[145] BREMMER 1987; UNGERN-STERNBERG 1993.

[146] See Varro, *Ling.* 6.19: *togata praetexta*.

[147] Cf. GRAF 1993, 1993a; WISEMANN 1995.

[148] Cf. for Ovid's *Libri fastorum* MILLER 1992; PORTE 1993; NEWLANDS 1995; GEE 2000; FANTHAM 2002; MILLER 2002; GREEN 2004; RÜPKE 2009b: 131–7.

[149] See ZANKER 1968; 1990: 201–15.

[150] TURNER 1967, 1982: 18–21; VANSINA 1985.

by antiquarians and theologians have little to offer by way of a general interpretation of the calendar as a system. In the surviving texts at least, the expected all-embracing experience of Roman religion, available, we were told, only in the written media, is by no means self-evidently present.[151] In what follows, therefore, I intend to return to an analysis of the *fasti* themselves.

Previous chapters have emphasized the potential inherent in the *dies natales* for establishing historical associations. Conditions in the Imperial Period and under the Republic are not strictly comparable in this respect. The *gens*-based structure of the Republican nobility gave history a powerful legitimizing role. This manifested itself, most particularly in respect of the later annalists, in a historiography dominated by 'party-political' concerns, to the extent of massive falsifications of history.[152] In view of this explosive element in social and political life, we are entitled to assume a rudimentary historical awareness, tending to be associated with such concrete, topographically prominent monuments as temples, on the part of a circle extending beyond the bounds of the real nobility.[153] These conditions, and the strategies associated with them, were rendered obsolete by the upheaval of the Caesarean-Augustan 'revolution', with its prosopographically demonstrable effect on individual lives and careers, and resultant revaluation of political institutions. The historical dimension of the temple-foundation days was reduced, in a process that was enormously promoted by the changes already described to the dates themselves. Thus it was above all the most recent history that came to be documented by the newly dedicated temples, as regards their presence in the calendar, if not their architectural reality and local continuity. It was, moreover, the most important dedications that were promoted to the status of *feriae*.

The historical dimension of the new holidays is impossible to overlook: it is made explicit in the *fasti* themselves. The historical event that provided the occasion for the constitutive Senate decree is mentioned with some regularity in most calendars; so far as the timeframe recorded is concerned, however, its location is not the distant past. It is, rather, the present or the immediate past that is addressed. The dates do not go beyond the forties, Caesar's seizure of power, and the beginning of the rise of the son C. Caesar adopted by Julius in his will.

The comprehensive nature of the information given in the *fasti* for the new *feriae* provides a helpful standard for the other dates, those of the traditional festivals. Only two instances include an additional historical note: already in the Republican *fasti*, the Parilia bear the additional note *Roma condita*; the *dies Alliensis*, the day of the defeat by the Gauls, is likewise already recorded in the Fasti Antiates maiores, although it is not associated with any cult activity.[154] Normally, the explanatory note added to the calendar entries for this group names only the patron

[151] BEARD 1991: 55–6 (with no supporting evidence).

[152] E.g. GUTBERLET 1985; RÜPKE 1992: 68–71.

[153] See Cic. *Fin.* 5.52.

[154] Not accounted for in BEARD 1987

divinity of the *feriae* mentioned after the letters *NP*. The extra information given by the Fasti Praenestini cannot be regarded as standard, but should be assigned entirely to the commentary component of this singular inscription.

What remains of our initial hypothesis? In view of the uncertainty surrounding the ritual explanations given by participants in the festivals (in Ovid's *fasti* commentary they are normally instructed, and not cited as authorities), no basis exists for ascribing a high degree of historical understanding to the public addressed by the *fasti*. In the case of Ovid, who is the most ready to locate an interpretation in Rome's early period, we should keep in mind the special circumstances of the Augustan period: in the context of a self-styled new beginning, legitimizing references to the early period are particularly significant.[155] In later years, the situation changed entirely: now, the Augustan period itself served as the founding past. Insofar as the calendar included the festivals of earlier emperors, and referred to their dates, it reacquired historical resonance, as is demonstrated by the Julian-Claudian dates of the Feriale Duranum from the Severan period. Here too, however, the emphasis is on celebrations of the current dynasty.

155 Scheid 1992: 124ff.; Nilsson 1951: 49–122.

9

The Disappearance of Marble Calendars

Prestige marble calendars are no longer attested with any certainty beyond the Early Principate. The end of the tradition is marked in Rome by the Fasti Pighiani from the time of Caligula, which, unfortunately, survive only in MS form, and, outside Rome, by the Fasti Sorrinenses minores, which are unlikely to be much later; both already show clearly deviant formal characteristics. The later, second-century Fasti Lanuvini and the *'fasti' Guidizzolenses* (which can scarcely any longer be described as such) are on another level as regards content and size.[1] On the other hand, the Fasti porticus at the end of the second century and the Fasti Filocali in the mid fourth century represent, respectively, painted wall calendars and book calendars that, in terms of their form, attest to a linear development from the Early Imperial-Period *fasti*, and, with their many contemporary emendations, exclude any idea of a 'classicizing' backward reference to that golden age of the *fasti*.[2] However, it would be surprising if the Mid and Late Imperial Periods had managed without a current calendar. Why, then, at a time when other epigraphic genres are developing positively, do the marble calendars disappear so early?

Before turning to other reasons for the disappearance of the marble calendars, it is worthwhile to re-emphasize a conclusion that emerged earlier from our analysis of individual calendars: while there are *fasti* that bear the 'official stamp' of a *collegium* or *municipium*, there is no sign of a central edition maintained by the *princeps*, the *pontifices*, or a dedicated calendar authority. An instance for comparison is China during the Qing Dynasty, where the calendar had outstanding significance as the most important symbol and instrument of legitimization of the emperor's cosmic role; here, the calendar for each new year was ceremo-

[1] RÜPKE 1995a: 149–51, 160–4.

[2] RÜPKE 1995a: 86–94.

The Roman Calendar from Numa to Constantine: Time, History and the Fasti, First Edition. Jörg Rüpke.

niously promulgated, and two million copies distributed to all public locations in the empire.[3] Of course, these figures for the sixteenth century and after have as a necessary condition the existence of appropriate printing technology; but they attest in particular to the degree of centralization.

It follows that the high degree of uniformity in the text of the calendar, and the insignificance of the existing variations, is one of the most striking traits of the Augustan-Tiberian calendar. A thought-experiment suffices to envisage the ease with which changes might have occurred: precisely because the *fasti* did not have to assume the practical function of a local *feriale*, nothing would really have stood in the way of the utmost variety of variations. Particular commentaries on the divinities important to the Imperial house, such as Apollo and Venus; the highlighting of gods worshipped in nearby temples; notes on days with particular relevance to associations between the emperor and the locality: all this would have been possible without excessive damage to the standard form, and virtually none of it actually occurred. A familiarity with the calendar, of the kind manifested by Ovid in his *fasti* commentary, would have been too much to expect of the producers of the *fasti* themselves: to give full and distinct treatment to dates of a similar type, so as to use the calendar as if it embodied a symbolic language, and, by rearranging its individual elements, or at least giving them a different emphasis, to arrive at new theological (and political) narratives. The very complexity of that symbolic language militates against its being used autonomously and innovatively: in other words, the threshold for altering the parameters of the genre is set so high, that only a Verrius Flaccus ventures to cross it by 'contaminating' text with commentary; to stress once again, Ovid wrote only a commentary. Faithfulness to the original was fundamental to any use of the calendar as a visual symbol in the *fasti* form.

If variety of expression was no longer available as a means of manipulating the calendar's content, only the material parameters remained to deliver competitive effect, and facilitate innovation and rivalry. Nevertheless, marble represented the only choice for prestige purposes, and decorative borders and in the final analysis size the only variables. The limit was reached in the late Augustan-Tiberian period, with dimensions reaching some 4.5 to 6.5 square metres (Fasti Verulani and Allifani). Here once again, it is the Fasti Praenestini, at about 12 square metres, that take the prize. Any further escalation in size would have required the covering of entire wall-surfaces. The 'queen of inscriptions', the Monumentum Ancyranum with the text of the *Res gestae divi Augusti*, shows where this might lead: the Latin text takes up both internal walls of the *pronaos*, in each case to a height of 2.70 metres and a width of 4.00 metres, while the Greek text covers the external wall in nineteen columns 1.25 metres in height, extending over more than 19 metres in all.[4] The Constantinian Forma urbis, together with its captions, is even many times

[3] Smith 1991: 74–5.

[4] Gagé 1935: 44–5.

this size. For the *fasti*, especially the calendar component, whose text could not be rearranged, or filled at will, a further increase in size could have been obtained only at the price of enormous empty spaces. Even the Fasti Praenestini, measuring 12 square metres with the extra text of their commentary, were unable to compete.

A further problem affecting the *fasti* arose not during but after installation: not only did the list of Imperial festivals constantly grow, it was also subject to frequent deletions. From the evidence of dated inscriptions, we may conclude that the major problem of the month-names was resolved at a more fundamental level by being ignored; that is to say, the old month-names continued in use. Although errors in the area of Imperial feast days were also frequent, numerous corrections and additions tell us that they were to be avoided, at least in principle. The succession of a new Imperial regime seldom occurred on a harmonious basis *vis-à-vis* its predecessor, and often led beyond the mere neglect of the earlier *princeps'* festival dates, to open antagonism against his institutions, to their restriction, and even at times to *damnatio memoriae.* Thus, already within the Julian-Claudian dynasty, Caligula and Claudius made inroads into their predecessors' *ferialia*, and the dynastic caesura after Nero's death provoked a new 'cleansing' on the part of the Senate.[5]

The sources show that personalities were not the only cause of such revisions; the restriction of state expenditure on games, and the removal of barriers to the judicial process, formed a motivation of at least equal rank.[6] The rise of Sejanus, entailing the public celebration of his birthday, and his abrupt fall, which resulted in the institution of *feriae* on the occasion of his execution,[7] illustrate the speed with which calendars had to be corrected if they were to keep up with the pace of events. Thus, above and beyond the one-off cost of endowment, constant maintenance was required: something that could be afforded by only a few, well-organized colleges, most prominently the Arvals, who, into the Claudian period at least, had their own *commentariensis* among their *servi publici*. This was perhaps a further possible reason for refraining from erecting an expensive inscription that might soon be obsolete.[8] But even the lists of *magistri* serving only small associations, their maintenance much more likely to be driven by special interests, easily became intermittent or ceased altogether, as is shown by the Fasti magistrorum vici.[9]

The problems indicated here must be seen against the background of a constantly developing epigraphic and artistic culture, offering a broad range of alternatives for an abundance of occasions and purposes.[10] Dedicatory inscriptions, honorific inscriptions, building inscriptions, not to forget epitaphs, statues, and reliefs, were all genres that offered individuals or groups the opportunity to give

[5] Suet. *Cal.* 23.1; Dio Cass. 60.17.1; Tac. *Hist.* 4.40.2; Dio Cass. 68.2.3; NICOLAI 1963: 201–2; TALBERT 1984: 386–91.

[6] Suet. *Aug.* 32.2; *Hist. Aug. Aur.* 10.10; Tac. *Ann.* 13.41.4–5.

[7] Tac. *Ann.* 6.25.2; *Inscr. It.* 13.2.257.

[8] MICHELS 1967: 142.

[9] RÜPKE 1998.

[10] ZANKER 1990; ALFÖLDY 1991.

expression to their social position, their cultural standing, religious adherence, and association with the ruling house.

These alternatives lead us back to where we began, with the lack of opportunities to vary content. The consul lists of the *fasti* had a Republican character, even in the Imperial Period: Horace is as able to use them in a poem addressing the ruler as to weave them into laudatory passages of one directed at an Aelius Lamia, later consul for 3 CE.[11] This was not possible with the calendar: Decimus Brutus did not receive his feast day, and no conceivable Aelius Lamia had any prospect of doing so. The more comprehensively annotated *feriae* were exclusively reserved for members of the Imperial House, and the slight prospect of following in the footsteps of the traitor Sejanus, and receiving one's own day, was scarcely one that a Roman would have seen as an honour worthy of his *dignitas*. Even the possibility of being commemorated indirectly, via the dedication date of a temple one had founded oneself, was of only slight import: it was the competent bearer of the *praenomen imperatoris*, or one of his closest relatives, who celebrated the triumph for the victorious campaign that brought the means to build the temple, and that was more likely to remain in the public consciousness than the statue of the army commander active in the field, with all its triumphal insignia.[12]

The monopolization of the calendar by the *princeps* had two serious disadvantages so far as honorific use of the *fasti* was concerned. Tacitus points out the first of these with his reference to *fastos adulatione temporum foedatos*:[13] *fasti* disfigured by the intemperate adulations of epochs gone by, to be removed at the first suitable opportunity. The same idea, in practical terms, is conveyed in the critique, expressed according to Tacitus by C. Cassius in 58 CE, warning against making every aspect of a victory – the day of the battle, the day of the news, the day of the report to the Senate, etc. – the occasion for a feast day:[14] too eagerly was the smallest military action or the slightest public or familial event seized upon by sycophants as a historic turning point, to be proclaimed *feriae* by the Senate. The salvation of the Republic – still, to Caesar and Cicero, the criterion for proclamation in calendar entries – had long since ceased to be the guiding principle. In the exposure occasioned by every impugnment of a predecessor by his successor, the practice put calendars themselves in a bad light, as documents of the *princeps*' overweening zeal, or of his discomfortingly exaggerated sense of his own importance: a perception that, under the last Julian–Claudian rulers, must have spread beyond the circle of those of the Senatorial elite who still thought of themselves as 'Republican'.

The second disadvantage was engendered by the social hierarchies in Imperial cities and towns. Whoever erected a calendar displaying the many important achievements of the ruling house did not do so in the expectation that the emperor

11 Hor. *Carm.* 4.14.4; 3.17.4; cf. FEENEY 2007: 170–82.
12 RÜPKE 1990a: 208. 233–4. 241.
13 Tac. *Hist.* 4.40.2.
14 Tac. *Ann.* 13.41.5.

would be aware of it, and of him as a loyal subject. The message was directed more at those who would probably read or at least see it, the patron's fellow citizens, or fellow members of his association. The Imperial cult itself, to which we should in the broadest sense assign the installation of a calendar, offered a medium that was apparently addressed to a common superior, in the form of an absent third party, and so could be all the more unpretentiously directed at the citizen and comrade at one's side: however questionable the purpose it might actually serve, the installation of a calendar may be interpreted as an example o of euergetism.[15] If the desired effect of any gift is to fall back to the giver, at least indirectly, in 'generalized reciprocity', above all in the form of prestige, designed to stabilize social position, and thus perpetuate social imbalance, the gift itself must provide for self-representation in some appropriate manner: a prominent position in the sponsored festive procession, or documentation of honorific offices held on the inscription recording the donation of food to the masses.

This is where inflexibility of content told against the marble *fasti*: they offered no formula that afforded adequate, more than marginal space for the patron's self-representation. The free space on the calendar was confined to the margins: captions that had no intrinsic connection with the unchanging text of the *fasti*, and could not be allowed to attain too large a size relative to the calendar area, which was perceived in the main as one sole visual entity. So it is no wonder that, to boast of erecting *fasti*, one chose a separate inscription rather than the *fasti* inscription itself.[16] This did nothing to promote the calendar as a form of 'investment'. And then the adulatory element: of the feast for Nero, even under Vespasian, the memory of the fat roast pigs and the wine remained, while the marble calendar with Nero's dates was simply embarrassing, whether because of the entries or because of the clearly visible deletions. Taking all this into consideration, here, to my mind, lie the strongest grounds for the disappearance of calendars after the Claudian period. Another problem to be considered is that of the multi-annual festivals, which grew in number under the Empire. Greek *ferialia* – an example is the Nicomachean inscription in Athens[17] – addressed the problem by having dedicated lists, separately combining the two-yearly and the four-yearly festivals;[18] the Roman *fasti* could not accommodate such dates.

And yet they lived on, as painted wall calendars, and as magnificent or – unfortunately non-extant – modest book calendars. There is no evidence at all of bronze calendars; although their existence, as that of wooden specimens, whose chances of survival would have been slighter still, is impossible to rule out a *priori*, two considerations allow us to refrain from pursuing any such possibility: firstly, the problems and functions connected with such artefacts would coincide very

[15] VEYNE 1990; PRICE 1984; QUASS 1993: 303–46.
[16] *CIL* 10.6679.
[17] DOW 1961, 1957.
[18] PARKE 1977: 14.

strongly with those pertaining to their stone counterparts; and, secondly, the evolution of the two extant forms – an evolution that would not be inherent to the last-named sub-genres – is such as to preclude the use of the materials mentioned. Of course, this train of reasoning relates primarily to prestige specimens. Simple types of wooden calendar may very well have been widespread; the marking of propitious days with chalk makes the use of this material more likely. The use of small stones (*lapilli*) for the same purpose would also be compatible with wooden calendars.[19] But a new development was in the offing. Both the Fasti porticus and the Fasti Filocali are illustrated calendars: decoration had moved from the margins to become the visually dominant element. The audience addressed also changed: these calendars spoke primarily to their owner, to the individual reader inside the building, within the closed portico; they did not hang on the external wall or in public spaces, did not address the chance passer-by, or the audience internal to a private club.

With the altered audience, the intention had also changed: veneration of the ruler no longer played a determining role. This is no mere assertion; it can be discerned in the graphic form of the monthly columns. The Imperial festivals may have grown in number in relation to the Early Principate, but, visually, they remain unremarkable. In terms of size of character and length of entry, they do not differ from other dates: *Tiberinalia* and *Augustales*, *n(atalis) Asclepi* and *n(atalis) Traiani*, stand side by side without differentiation on the decorative pages of Furius Filocalus. Calendars, *fasti*, existed under Augustus as under Constantius II; but their most magnificent manifestations served entirely different purposes in each case.

[19] Hor. *Carm.* 1.36.10; e.g. Mart. 9.52.5.

10

Calendar Monopoly and Competition between Calendars

10.1 One Calendar

Having remained constant for centuries, the face of the *fasti* changed within the space of the few decades that began with Caesar's dictatorship. Cult reflections of current events – victories of the *princeps*, and offices assumed by him – enter in great number into graphic representations of the calendar. In this, as our analysis has already shown, the *fasti* did not function merely as a mirror of religious politics. Rather, religious politics themselves often took on such a form as enabled them to find, in papyrus-roll, wall, and stone calendars, a kind of autonomous expression that could no longer be accomplished within the common space of ritual activity. Such a process is entirely usual today, in the changed conditions of mass communication: the scenarios and timetables of political events are managed with a view to achieving optimal representation in the media. What occurred under the Augustan watchword of *pietas* was the sacralization of a medium that, while it did not itself regulate sacral activities, had been capable of being interpreted in this sense ever since Fulvius' inclusion of temple dedications.

Two factors share particular responsibility for the circumstance that the *fasti*, designed to serve primarily political-juridical purposes, and functioning as a calendar of record, became increasingly important as a document for displaying religious facts. On the one hand, the new *feriae* played an important role. The relationship that had applied in the Republican *fasti* between the daily character as the crucial item of information and the festival abbreviation as justification of that character was reversed: although daily characters, in this case *N(efas) P(iaculum!)*, at first continue to be entered correctly, the circumstances defining them – with the exception of the Augustalia, which even appear in abbreviated form on 23 September – are no longer reduced to a flexible name capable of abbreviation. No longer is the political-juridical norm explained (*NP, quod feriae*

The Roman Calendar from Numa to Constantine: Time, History and the Fasti, First Edition. Jörg Rüpke.

...), but the religious factor underlying that norm (*NP. Feriae, quod ...*): validation, and thus the message of the *fasti*, has been moved one stage further back. That, in the process, the theological message – whose *feriae* are they? – gives place to the historical message – why were these *feriae* established? – represents an internal problem at the religious level, and does not affect the structure outlined.

The exhaustive manner in which the religious elements are validated is all the more significant insofar as there is reason to suspect that the religious information provided here through the medium of the emperor or the Imperial house is a non-random selection, not to be found in this form in any other document. The great number of a specific type of *feriae*, their differentiation from other memorial days (mainly known from the commentaries of the Fasti Praenestini, and lacking festival status), and, finally, the constant revisions, contribute to this non-random character, this systematic, conscious selectivity that emphasizes particular biographical, familial, and constitutional aspects of the Principate and the *princeps*. The point is highlighted during the Imperial Period by the quasi-authoritarian recruitment of the great priesthoods, expressed, broadly speaking, in their universal participation in all (new) rituals. What we have in the *fasti* is not a chance collection of highly differentiated rituals performed by different groups of religious specialists, as the ensemble of ancient *feriae* might be characterized, but a catalogue of dates whose cult character is largely uniform, and that are scarcely ever specifically assigned.[1]

The second factor comprises the disappearance of the daily characters. The evidence of the Fasti Sorrinenses minores, which still provide daily characters, but with incipient signs of their decline, and the Fasti Lanuvini, in which those characters are no longer present, suggests that they disappeared between the end of the first and the beginning of the second century of the Christian Era.[2] Unfortunately, the dates of both inscriptions are too uncertain to provide a basis for establishing a chronology of the process without descent into a circular argument. For a safe *terminus ante quem*, we have to wait for the jurist Gaius; writing in the Antonine period, he speaks of the *dies nefasti* as of a long-defunct institution, requiring explanation.[3] The disappearance of this eponymous element of the *fasti* entailed the end, in visual terms, of the calendar's time-management potential in respect of the juridical system: although, in fact, this function certainly continued, from now on the requisite data had to be obtained from the religious entries. No longer was the religious element refracted by the juridical information; now the converse was true. Increasingly, the classical legal texts speak of *dies feriatici* or *feriae* instead of *dies nefasti*.[4]

In view of the absence of some data of a political-juridical nature in the earlier period too, including the Republican Period, our portrayal of this development

1 See *Inscr. It.* 13.2.115 (Fasti Praenestini 17. gl.); Lucan, 1.592–604; Tac. *Ann.* 3.64.3–4; Feriale Cumanum.
2 Rüpke 1995a: 145–51.
3 Gai. *Inst.* 4.29.
4 Ulp. *Dig.* 2.12.2; *Cod. Iust.* 3.12. Cf. the untechnical usage of Plin. *Ep.* 10.8.3.

must be qualified. We learn from a letter from Cicero to his brother Quintus that the Saturnalia marked the end of the Senate term.[5] The phraseology does not permit the inference of hard and fast rules of sacral law, but it does suggest an established practice.[6] The holidays for the harvests of grain and grape provide a much weightier example. Since the discovery of the *lex Irnitana*, it must be apparent that, as early as the Flavian period in Rome, a length of time amounting to two months was kept strictly free of hearings and sittings.[7] If Augustus himself laid down rules for the dismissal of the Senate in September and October,[8] the institution must be older still. The facts tell the same story: the problem was certainly not new to the Imperial Period. Presumably, an appropriate break was announced by the *praetor*, and, once the two months in question had become established, it was incorporated into the *edictum perpetuum*.[9] The facts of the matter as described in respect of Augustus, and in Spanish municipal law, eluded the terminology of the *fasti*. This terminologically and therefore in a certain sense also legally defective determination may be partly responsible for the deficient tradition from the Republican Period. The wording is still remarkably laborious in the document from Irni, when it speaks of 'days on which matters are deferred'. Only with the demise of the terminology of the Flavian *fasti* was the way opened for old words to be reused in more concise formulations.

By the nature of the historical origins of the daily characters, their disappearance affected not only the juridical system, but political activity too, in the form of the *ius cum populo agendi* exercised in the *comitia*. The restriction of the *dies comitiales*, with its aim of minimizing the element of political uncertainty inherent in the people's assemblies, was one of the most important tactics of political confrontation during the final years of the Republic. The repetition of rituals (*instauratio*) or the imposition of *feriae imperativae* already forestalled the holding of *comitia*, which, once begun, could be halted only by the augural emergency brake of *obnuntiatio*, the announcement of non-auspicious signs. With the sole rule of Augustus, the people's assembly largely lost its potential for unrest: not so much because a once active body had fallen into passivity, as on account of the concerted action of the magistrates, who previously constituted the polarizing element. The people's assemblies, the *comitia centuriata*, continued to meet in order to elect magistrates and adopt laws (*leges*). Predetermined institutional steps, some of them new – recommendation of the *princeps*, decision of the Senate, elective decisions of special, honorary *centuriae* – meant that the balloting of the *comitia* became *de facto* a matter of formal acclamation.[10] From the point of view

[5] Cic. *Q Fr.* 2.1.1–3. P. STEIN (1930: 119) registers only one exception during 68–43: 20 December, 44 BCE.

[6] Cf. Suet. *Cal.* 6.2; *Lex de Germ. tab. Siar.* 2 b 3 (LEBEK 1993: 88–9); Tac. *Hist.* 3.83.3 (3.78.1).

[7] Against Ulp. *Dig.* 2.12.pr.; *Hist. Aug. Aur.* 10.10; now *Lex Irnit.* 5 C 24–45 clearly referring to older urban institutions.

[8] Suet. *Aug.* 35.3; TALBERT 1984: 200, 211.

[9] Cf. *Dig.* 2.12.4

[10] FREI-STOLBA 1967.

of the average citizen, however, who, even under Republican electoral law, was seldom called upon to cast his vote, little may have changed.[11]

Under such conditions, management of the calendar in respect of the *comitia* was of no great import. The few, uncontroversial meetings each year could be restricted by conventional practice to a small number of 'permitted' dates, and there was no need to make full use of the time constraints that had been indicated by the *fasti*, or even to continue to publicize them. If this meant that, in many areas, traditional augural divination fell into disuse, it attests *ex negativo* to the eminently political function and specific objectives of this mode of divination and its personnel. The only committee that still had any relevance was the Senate, military revolts and other political activities of non-political institutions never yet having been susceptible to regulation by the calendar.

The situation during the major part of the period of the Republic was clear: the Senate could meet whenever it wanted. Perhaps there were particular days that were avoided on religious or other grounds; but investigation of all known sitting dates shows that such exclusions, if they in fact conformed to any system, did not coincide with any category indicated in the *fasti*.[12] Practically speaking, this entire freedom of disposition in respect of time, restricted only as regards choice of place, dictated by the use of particular *templa*,[13] corresponds with the Senate's central function in the Republic's political system.

There were practical grounds for avoiding clashes with meetings of the people's assemblies, which were conducted by magistrates who also had a presence in the Senate.[14] The avoidance of such clashes, and of use of the *dies comitiales* at all, excepting the case when a people's assembly had already finished,[15] was given statutory force by the *lex Pupia*.[16] The law's initiator, unless we look to the first half of the second century,[17] which would be quite unlikely for such massive regulation of the Senate, must be identified as M. Pupius M. f. Piso Frugi Calpurnianus, and its passage assumed for the year of his praetorship (72 or 71 BCE) or his consulate (61).[18] This would place it chronologically near to the *lex Gabinia* of 67, which charged the February Senate sittings with the handling of embassies.[19]

11 JACQUES, SCHEID 1990: 49–52.

12 Unfortunately, Gellius (14.7.9) tells us only that Varro, in his *Commentarium eisagogicon*, talked about *quibus diebus haberi senatum ius non sit*. The subject of Varro's remark remains unclear; it must have concerned individual existing statutes on this subject. The use of the term *ius* indicates an intentional departure from the terminology of the *fasti*. BONNEFOND-COUDRY (1989: 199–229) has fully appraised the dates, and drawn the conclusion recorded here (220). Cf. P. STEIN 1930: 119.

13 Varro *ap*. Gell. 14.7.7.

14 MOMMSEN 1888: 921–2.

15 Dio Cass. 37.43.3; Cic. *Att.* 1.14.5.

16 BONNEFOND-COUDRY 1989: 221.

17 Thus MOMMSEN 1888: 923: *c*.154 BCE.

18 A safe *terminus ante quem*: Cic. *Sest.* 74 (57 BCE). Cf. for February, 61 BCE Cic. *Att.* 1.14.5.

19 ROTONDI 1912: 373. See Cic. *Fam.* 1.4.1; January is given prominence (Cic. *Q. Fr.* 2.2.3; Caes. *B Civ.* 1.5.4), but not addressed in isolation: Cic. *Fam.* 8.8.5; *Att.* 1.14.5.

The content of the *lex Pupia* was constrained by the fact that any mechanism, whether it be the eight-day nundinal cycle or *feriae imperativae*, that barred regular *dies comitiales* to *comitia* must by the same token have made them available for Senate sittings.[20] Furthermore, while specifically directed against decision-making Senate assemblies, the law appears to have left the way open for consultative meetings and particular subjects: a *senatusconsultum* of 51 BCE expressly permits discussion of the question covered by the decision on *dies comitiales*, while also allowing in this case the possibility of Senate resolutions.[21] Only thus, too, can we explain the increase in *dies comitiales* among dates for meetings of the Senate in the period after 63 BCE.[22]

That the ruling should not be seen as an attempt to establish clear, calendar-based jurisdictions is attested, despite its revolutionary author, by a plebiscitary *lex Clodia* of 58 BCE, now opening up *dies fasti* too to *comitia*.[23] The separation established by the *lex Hortensia* between *dies fasti* and *dies comitiales* was thus nullified; paradoxically, only the Senate was now still affected by it.

Under the new system, there was no reason to restore the situation where the Senate as central organ of the Republic was not, in principle, subject to any restriction in choosing the dates of its sittings. Augustus himself appointed two fixed days per month for sittings, when a *senatus legitimus* was automatically summoned:[24] although this did not mean that the Senate's activities were restricted to those days. The critical factor is not the altered number of sittings, but the very fact of regulation.[25] The calendar presumably continued to include these days in the Claudian period; the Fasti Lanuvini still have them alongside the daily characters, and in later calendars they feature in isolation. This shows that the function of the new *fasti*, to serve, like the old, as an instrument of political regulation, had persisted: it was this calendar, if not still, then once again, that determined the political system's 'opening times'. The fact that the Senate, previously the instigating agent of such regulations, was now primarily their object, is indicative of important changes to political structures and parameters. The point of departure of this revolution in the power of disposition over time may be seen in the discussion of the treatment to be given to the anniversary of the day of Caesar's assassination.[26] It should be noted that the incompatibility between festivals and Senate sittings remained valid in the later Imperial Period: there is a convergence between religious and political rules.[27]

20 MOMMSEN 1888: 923.

21 Cic. *Fam.* 8.8.5; MOMMSEN 1888: 923.

22 Noticed by BONNEFOND-COUDRY 1989: 223.

23 Cic. *Sest.* 33. 56; THOMMEN 1989: 243.

24 Suet. *Aug.* 35.3; Dio Cass. 55.3 (9 BCE). The change is accompanied by ritualization (TALBERT 1984: 224–5).

25 See Dio Cass. 58.21.2.

26 Suet. *Iul.* 88: *Curiam, in qua occisus est, obstrui placuit idusque Martias Parricidium nominari, ac ne umquam eo die senatus ageretur.* According to the Fasti Filocali, the mid-March *senatus legitimus* did not take place, as in most other months, on the Ides, but *pridie idus* (*Inscr. It.* 13.2.243).

27 *Hist. Aug. Claud.* 4.2; cf. Tac. *Ann.* 3.23.1.

Some isolated developments are indicative of a tendency to ascribe comprehensive regulative authority to the calendar represented by the *fasti*. This process appears with particular clarity in the religious sphere, which the Republican *fasti*, especially in their original form, had largely left aside, only claiming here to exercise the function of a calendar of record, over and above their specific management of time in the area of politics and the law. Three phenomena attest to the new claim.

In 30 BCE, in the aftermath of the victory at Actium and Antonius' final defeat by the capture of Alexandria, the Senate declared 14 January, M. Antonius' birthday,[28] *dies vitiosus*.[29] The intention behind this act is beyond doubt: the day was qualified a day of misfortune. This is our first (and last) encounter with the concept *dies vitiosus*. Nowhere do we learn what consequences were implied by it. *Vitium* in augural law denotes an error that renders an act, such as an election, invalid.[30] But we know of no augural rule that would provide justification for such a categorization of this day. Once only, in a letter of Cicero in 55, is there mention of *dies vitiandi*, and here, significantly, it is the people's tribunes who let it be known, by way of rumour, that they wish to use this obstructive measure. We are left in ignorance as to what it meant in practice, and whether Cicero is really using the term as *terminus technicus*.[31] In the absence of a clear understanding of the words on the part of the augurs, the senators may have had the *dies religiosi* in mind, days upon which religiously motivated prudence (*religio*) advised refraining as far as possible not only from sacral actions, but from any other kind of activity.

Dies religiosi as such are not indicated in the calendar, although they are determined on its basis. There is, however, an analogous instance in the *dies Alliensis* on 18 July; this belongs to the category of *dies religiosi*, and, in commemorating the Gaulish catastrophe, certainly marks a day of great misfortune.[32] But it must be noted that even this day is not designated *dies religiosus*, at least in the *fasti*; probably one of Fulvius' entries, it relates to the only other historical note in the Republican *fasti*, the foundation of Rome on the Parilia.

Despite its well-defined motivation, this initiative by the senators is something entirely new. Drawing for aid upon the augural metaphor of *vitium*, for a single day they created and introduced into the *fasti* a category of days that, legitimately, could have been so constituted only by the religious bodies in the form of a declaration of *religio*. This balancing act, which marks the instrumental intrusion of the *fasti* into the realm of the religious calendar, remained without direct consequence. Although Claudius does not appear to have rescinded it (this singularity

[28] For the pre-Julian nonexistence of *a. d. XIX Kal. Febr.* see Ch. 6.2, 6.3 and SUERBAUM 1980; audaciously interpreted by FEENEY 2007: 155–6.

[29] Dio 51.19.3; *Inscr. It.* 13.2.72.

[30] LINDERSKI 1986: 2159. 2162–77.

[31] Cic. *Att.* 4.9.1

[32] Liv. 6.1.11–12.

was evidently something of a hot potato), he did mitigate its profile by expressly heightening that of the celebrations for his father Drusus on the same date.[33]

The condemnation of this day had at least an indirect consequence in the handling of the *dies natalis* of Agrippina on 6 November; according to Tacitus, after her murder by Nero in 59 CE the Senate ruled that the day *inter nefastos esset.*[34] In practical terms, this is not impossible; in earlier calendars the day is marked *F(as)*, in common with most of the days following the Nones, and this qualification could be changed. As a day following a feast day (*dies postriduanus*), however, it was already *dies religiosus.*[35] We know of no further use of the concept *dies vitiosus*. It had not been employed for ninety years, and the only instance may have gradually disappeared from calendars after Claudius; as 14 January is marked *E(ndoitio Exitio) N(efas)*, and therefore was *fastus*, and so remained in its central portion, the expression *inter nefastos esset* as a description of the event of 30 BCE would be mistaken.

Although the simple alteration of a *dies fastus* to a *dies nefastus* would have had (intended) consequences similar to those of the designation *vitiosus*, there is no precedent for such a measure either, no antecedent among the instruments available to the calendar, unless we assume wholesale terminological dissolution, at a time when the system of daily characters was itself beginning to break up. This may have been a contributing factor, but it must remain an extravagant hypothesis.

The simpler assumption is that the day was regarded as *inter nefastos* in the sense of being classified an *NP* day: as we saw in the context of our discussion of Fulvius' categories, this differentiation is consistently neglected in literary sources. The intended sense would then have been that *feriae* were established on this otherwise intrinsically unremarkable day: an entirely normal step, for which may be cited by way of example not only the *feriae denicales*, to be observed on the occasion of a funeral,[36] but the permanent *feriae* instituted for the execution of Sejanus and the death of Agrippina.

A second factor, a consequence of the intrusion into the religious sphere, left much deeper traces in the *fasti*. Because the one, 'state-approved' calendar increasingly claimed the status of a religious norm, it had also to be descriptively correct.[37] I see in this the reason why the Late Antique Fasti Filocali included in its calendar Attis and Cybele (Mater Deum Magna Idaea), along with their most important festivals.[38] Of course, the mere statistical popularity of a festival was not sufficient reason to include it in the calendar: official, and in every case that means Imperial approval, had also to be given.[39] Critical, however, for inclusion in the *fasti* –

33 Suet. *Claud.* 11.3.

34 Tac. *Ann.* 14.12.1.

35 See Fasti Praenestini 14 January, *Inscr. It.* 13.2.113.

36 WISSOWA 1912: 433.

37 Cf. GIESECKE 1991: 240 for the Early Modern Period.

38 *Inscr. It.* 13.2.243.

39 SALZMAN 1990: 164.

festivals of Isis are to be found in the *menologia rustica* as early as the first century[40] – was the altered conception of the *fasti*. Isidis navigium, Arbor intrat, Sanguem, Hilaria, Requietio, Lavatio are neither all *feriae* nor all *dies natalis* of corresponding temples: religious events were documented, as earlier in the case of games (but only as an *addendum*), and this happened at the same informational level as everything else. No official 'Christianization of sacred time' is yet to be discerned here.[41]

Finally, and our third point: the religious significance accumulated by the Imperial *fasti* came to be reflected in the religious practice of broad sections of the population. Analysis of the dated dedicatory inscriptions of the Imperial Period,[42] and especially the extant festival lists of professional colleges,[43] shows that the Imperial festivals defined the agenda of individual cult practice to a greater extent than did the traditional *feriae*. Where an association's *feriale* represents nothing more than an extract from the *fasti*, there is in reality no more than *one* calendar.

10.2 Coexisting and Competing Developments

Fifty years after the Fasti Filocali, the lavish codex calendar of 354, little remained of the substance of the festival entries. The withdrawal of the financial infrastructure, beginning in 364 and decisively accelerated by Gratian's laws of 382,[44] the repeated occasions upon which sacrifices, as the most important sacral act, were banned,[45] and the Emperor's rejection of the supreme pontificate,[46] combined to break the organizational backbone of traditional state religion, with repercussions for many cults associated with it.[47] Massive suppression must not be confused with a quick death by the cessation of life-support systems: the correspondence of *pontifex* Symmachus demonstrates that this particular college continued to be organized and extensively active at the end of the fourth century.[48] But the interventions had direct consequences for the calendar. With their change in status, most rituals of these cults lost their claim to a place in the *fasti*.

It is beyond any doubt that a number of popular festivals continued to be celebrated alongside the festivals of the Imperial house, which were themselves

40 *Inscr. It.* 13.2.287 (*Menol. Colot.* Mart.), 293 (*Menol. Vall.* Mart.).

41 For an alternative view: SALZMAN 1999.

42 HERZ 1975.

43 AUSBÜTTEL 1982: 33–4.

44 364: *Cod. Theod.* 5.13.3; 10.1.8. 382 CE: Symmachus, *Relat.* 3.11ff.; Ambrosius, *Ep.* 72.14. 73, *Extr. coll.* 10.2. 6. 10 (*CSEL* 82.3). METZLER 1981; CANCIK 1986: 72; SALZMAN 1992.

45 NOETHLICHS 1971, 1986.

46 RÜPKE 2008: 63–4.

47 CHUVIN 1990; CAMERON 1990.

48 Symmachus, *Ep.* 6.29; 9.108. 147–8. See VAN HAEPEREN 2002: 202–11.

either actually or potentially religiously neutral. The removal of rituals with a heavy theological burden, such as the sacrifice and the *pompa circensis* of figures of the gods, and the downplaying of theological exegeses in respect of rituals such as the Lupercalia, allowed such practices to continue, on a basis that ranged from active reconfiguration and continued patronage to toleration.[49] In isolated instances in the fifth century, when it came to placing Church festivals on the calendar, conflicts with pagan rituals were still being consciously chosen; as general mechanisms, we can discern adoption and continuation as well as polemic contestation and avoidance. The 'Birth of Our Lord', the Christmas Festival, appears to have preceded the Feast of Sol Invictus.[50]

This dense process of substitution was not accompanied by any corresponding transformation in the *fasti*. On the one hand, the technical context remained the same in some degree of detail. Despite competition from the Greek and Oriental practice of counting the days serially through the month, this did not supplant dating by the Nones, Ides, and Kalends until the Late Middle Ages;[51] the calendar's configuration in columns, and the regular succession of red and black, characterizes Christian liturgical calendars to this day; apart from some minimal modification to the mode of intercalation, which can be discerned on three occasions in four hundred years, not even the Gregorian reform changed technical parameters of the calendar such as the names and lengths of the months. However, despite graphic correspondence between, say, the Fasti Filocali and medieval calendars, there is no broad continuity as regards nature and function of content (thus justifying the terminological distinction). The *fasti* of Silvius Polemius, from the mid fifth century, are the clearest case in point: while formally adhering to the structure of Imperial-Period *fasti*, so far as content of the daily entries is concerned they reveal themselves to be an antiquarian hotchpotch, a deviation that can be explained only by the lack of a binding contemporary model.[52] The most important developments appear to occur no longer within the 'one' calendar, but alongside it and in competition with it. These processes during the final portion of the historical phase ought not to be treated as the beginnings of the Christian 'calendar', but as a context of the Imperial-Period *fasti*, and a possible factor in the genre's demise.

The Republican *fasti* had never claimed to document all cult events, or even the great cult events. It was the juridical-political consequences of a day's religious status that determined its inclusion: if there were no such consequences, there was no entry. Many rituals were listed only in the *ferialia* of individual priesthoods or temples; multi-day *feriae*, as in the case of the Quinquatrus or the Saturnalia, went unremarked. At the same time, entire cults were excluded. The cult of Isis, which

[49] HOLLEMAN 1976; DUVAL 1977; ULF 1982; *Coll. Avell.* 100 (*Sources Chrétiennes* 65, 162–88).

[50] FÖRSTER 2007; RÜPKE 2006b.

[51] HENNIG 1980: 14.

[52] RÜPKE 1995a: 151–60; DULABAHN 1987.

already enjoyed a considerable number of adherents in the heart of Rome during the last years of the Republic, saw its festival dates go unnoted in the *fasti* into the late Tiberian period. Instead, it was exposed to persecutions and expulsions.[53] Subsequently too, only a small portion of its festival dates was recorded in the *fasti*. For this reason, in order to manage its everyday activities and ritual high points, the cult probably had constant need of a *feriale*, which surely must have been in writing, if one thinks of the advanced state of Egyptian temple administration.[54]

The situation was analogous for many other cults. Even if we concede the unifying tendency of the Imperial-Period *fasti*, it is inconceivable that a single calendar could have satisfied all Rome's documentary requirements. There remained a need for a great number of *ferialia* below the level of the *fasti*. The question that now arises is, did all these 'calendars', belonging to various groups and institutions, operate within a calendar structure placed on definitive display in the *fasti*?

There are two parts to the answer to this question. First, we can point to the unambiguous testimony of the sources: inscriptions from the city of Rome contain no trace of an alternative calendar identifiable by month-names or by a mode of counting the days that differs from the characteristic system based on the Roman orientation days. Similarly, there is no corresponding trace in the *corpora* of inscriptions belonging to groups who, by their origin, language, or religious attachment, might be judged most likely to have possessed such a calendar; this applies both to the now updated *Corpus Inscriptionum Judaicarum* and to the *Inscriptiones Graecae Urbis Romae*.

To the second part of our answer: what force should be attached to this *argumentum e silentio*? In respect of the Jews at least, we have positive proof that their religion was also practised in the Diaspora, and that it possessed a festival calendar that was authoritative, with modifications at least, beyond the Palestinian region of origin. The fact that these festival dates were superimposed onto the lunisolar calendar of Palestine, which was in no way inferior to most other Mediterranean calendars, suggests that this calendar also applied to the cult in the Diaspora. But the epigraphic evidence does not confirm this general observation in the case of personal dates. The few dates in Jewish inscriptions and graffiti in the city of Rome are exclusively Roman.[55] As the closed burial areas suggest an audience that was confined to the group rather than embracing a more general public, here too the Roman calendar is shown to constitute a practical dating system that extended to the intimate sphere of the date of death. It was against this very degree of integration[56] that the Imperial-Period 'lunarization' of the Jewish calendar was directed.[57]

[53] MORA 1990: 72–112.

[54] OTTO 1905–8; ARNOLD 1992: 13–57; EVANS 1961; QUAEGEBEUR 1979; *feriale* of Tebtunis *P. Oslo.* 3.77.

[55] *CIJ* 2, 68, 127, 271, 482, 528, perhaps 202, unclear 379.

[56] POLIAKOFF 1989; FELDMAN 1993; TREBILCO 1991: 142–4, 186–8; cf. MEINZER 1992: 135–40 for a Christian diary under the French Revolutionary Calendar.

[57] STERN 2001.

Christian inscriptions produce the same result.[58] Its significance is quickly confirmed if we turn our gaze to the world beyond the confines of Rome itself. The evidence, attested in inscriptions everywhere, is most easily perceived in the Acts of the Martyrs: the dating system used is a function of the particular locality; local practice is central; for the sake of completeness, perhaps, synchrony with the Roman calendar is at established.[59] Despite the religious prehistory of the month-names, there is no reluctance to use them, no polemic against them.

10.3 Eras

On the other hand, sensitivities existed in respect of the method of numbering years. Besides the multitude of cases where use was simply made of the locally prevailing system of 'dating' (or rather, of expressing reverence and assigning responsibility) based on eponymous officials or reigns, we find isolated instances of a critique against the political and socio-theological implications of such dates: to the period of office of the particular official or emperor is opposed the reign of Jesus Christ, in formal synchronism, but, in terms of content, as is shown by the use of particles expressing opposition, an unmasking of the illusory nature of the authority manifested in the dating function.[60]

It is in the continuation of this line of development that we find the genesis of the Incarnation Era, dating 'after the birth of Christ', with which Dionysius Exiguus replaced the era of the Christian persecutor Diocletian in 525.[61] This factor must, however, be seen in the correct light. The primary impetus for a new system of reckoning by eras lay in the need to consider periods stretching ever further into both the past and the future, demanded by the calculation of Easter cycles.[62] The first stage to be noted in the further progress towards implementation of the era system is the replacement of consular dating by dating according to the eras of Christian rulers (emperors and subsequently popes).[63] The consulates do not end until 534 in the West and 541 in the East.[64] The Incarnation Era was not finally implemented until Bede's chronographical and chronological works. An example from the 'opposing camp' reveals that the critique implied by the Incarnation Era against the prevailing practice was not an isolated phenomenon: in the *Vita* of the philosopher Proclus, written at the end of the fifth century, his death is dated

[58] NORDBERG 1963.

[59] *Mart. Polykarp.* 21, p. 18.4–8 Musurillo; *Mart. Apoll.* 47, p. 104.1–3 Musurillo and *Acta Pioni* 23, p. 164.30–166.4 Musurillo (cf. 2, p. 136.12 Musurillo).

[60] See Feeney 2007: 139–42, for a fine analysis of the pragmatics of eras.

[61] KRUSCH 1938.

[62] KRUSCH 1938: 4–10; STROBEL 1977: 138.

[63] *Novell. Iust.* 47.

[64] DEMANDT 1989: 200.

according to a 'Julian Era', calculated from the accession of Julianus ('Apostata').[65] The 'retrospective' Incarnation Era, i.e. years calculated 'Before Christ', represents a unique instance in cultural history. It appears for the first time in systematic use *c.*1292 in the anonymous Flores temporum, and only gradually established itself during the Early Modern Period.[66]

10.4 The Calculation of Easter

The demonstrable alternative eras have distracted us from the problem of the non-demonstrable alternative calendar. The beginnings of a rapprochement are offered by a date that is to this day, in the Gregorian year, determined on the basis of a different calendar: the Easter festival. Put crudely, Easter Sunday is the Sunday after the 'first' full moon in 'spring'. Thus, against the background of a purely solar calendar, we determine a festival on the basis of a 'lunisolar' calendar ('lunar' on account of the moon, 'solar' on account of the season), to which we additionally apply a sequential system of weeks. Although this calendar is in current use, it is purely theoretical; its antecedent is the Jewish calendar.

We can only touch upon the problems of calculation, enough to convey at least an impression of the complexity of the task, and awaken understanding of the centuries of dispute over the date of Easter. We shall leave aside entirely the problem of observation, although, recently, every calculation assumes empirical dates; there are, nevertheless, strong indications that even the lunisolar Jewish calendars, with their emphasis on the empirical method, used a cyclic system, probably of nineteen years.[67] The most serious problems have to do with definitions. When exactly is spring? What does 'after' the full moon mean, if the full moon falls on a Sunday, and that Sunday may begin at sunset (which, again, needs precise definition) on the previous day? Do we, even in the extreme case, wish to exploit to the full the limits of variability capable of being accommodated by a period of thirty-five or thirty-six days (one lunar cycle plus a week)? What does 'full moon' actually mean? To what locality should it apply?[68]

The only problem of interest to us here, however, is of a fundamental rather than a technical nature: how can the determination of a festival on a lunisolar basis be implemented in the context of a solar calendar? The common historical explanation states, in much simplified form, that the association in the calendar between the death of Jesus and the Feast of the Passover, combined with the

[65] Marinus, *Vita Procli* 36.

[66] VON DEN BRINCKEN 1979: 2–3.

[67] STROBEL 1977: 430–7; ZION WACHOLDER AND WEISBERG 1971. See also MULLER AND STEPHENSON 1974. Cf. STERN 2001: 47–85 for the empirical procedure.

[68] NEWTON 1972: 15–41; PHILIP 1921: 61–82; MAYR 1955; ZELZER 1978: 188–9, 196–204. See MOSSHAMMER 2008: 44–5 for the changing starting points of the Jewish 'day'.

Jewish character of the earliest phase of Christianity, makes it understandable that the corresponding ritual commemoration also continues to be attached to the Passover as defined according to the Jewish lunisolar calendar. While this practice continues to be recorded in Jewish-Christian circles into the fourth century, in the course of an anti-Judaism that is also observable elsewhere it gradually begins to seem suspect: Easter and Passover, like the Sabbath and Sunday, are differentiated in the calendar; the shift from the full moon, which, as 14 Nisan, in a month beginning with the observation of the first crescent moon, defines the Feast of the Passover, to the Sunday after full moon (even when the full moon falls on a Sunday) guarantees the separation of the two dates. The Council of Nicaea established this position definitively, and condemned the adherents of the practice of continuing to celebrate Easter on 14 Nisan, the date of the Jewish Passover (hence their designation as 'Quartodecimans').

Accurate as the picture drawn here may be for the period after Nicaea, it belies the manifold developments of the early period, which have been reconstructed by AUGUST STROBEL in a pioneering, if often hypothetical work.[69] It is precisely this early phase that is interesting from the point of view of the encounter between two calendars. STROBEL's conclusions put in doubt the idea of a Jewish calendar that was also of universal validity in the Diaspora, and served as a vehicle for conveying the lunisolar calculation of Easter to areas with solar calendars. It is in this light that I shall now give a brief account of those conclusions.[70]

The argument relies on the premise that, so far as the calendar was concerned, Judaism did not constitute a monolithic bloc. This especially applied to the Diaspora.[71] On technical grounds, and in view of the means of communication available, a procedure involving empirical observation of the first crescent of the moon, its promulgation in Jerusalem, implementation in the calendar, and subsequent publication, was not practicable for the entire Mediterranean area. Either decentralized observations were made and decisions arrived at in respect of the beginning of the new month, or there had to be recourse to calculating aids, and cycles. The inevitable consequence is that, even in those places where the 'Jewish calendar' applied, different dates emerged, which was of no importance in the normal run of events, that is to say as long as a state of isolation persisted. The situation was then further exacerbated by the existence of different Jewish calendar systems.[72]

This was the context into which Christianity evolved. From the beginning, it is possible to discern a marked interest in historical exactitude insofar as keeping the

[69] STROBEL 1977, 1984; cf. TALLEY 1982; HARNONCOURT 1985.

[70] It cannot be denied that STROBEL's complex argumentation is often compelled to remain in the realm of the hypothetical. The dates and attributions of important sources are uncertain, and calendrical systems have often enough to be extracted from the polemical remarks of opponents. The study is nevertheless given convincing substance by convergences between mutually independent observations, and the coherence of the overall body of material. But cf. now Mosshammer 2008 for details.

[71] STROBEL 1977: 355.

[72] Cf. now STERN 2003.

date of the Passover was concerned, not the least of motivations being eschatological concerns.[73] Despite later uncertainties, there was in this connection a highly confident tradition as to the true, Julian date of Jesus' death: 7 April, 30 CE.[74] The first point to be made, however, is that 'exactitude' is measured less by the local calendar at the time of the event to be commemorated, than by the calendar of the commemorating group, and depends on the method of conversion by which the date in question is mediated. The question as to what is identified as essential in the definition of a foreign local date, or an ancient date, has been sufficiently illustrated by the example of the Julian calendar reform, with its vacillation between continuing to reckon by the interval to the next Kalends and opting for the interval to the preceding Ides.

Reconstruction of ancient practice regarding the Easter Festival primarily reveals Quartodeciman, or more precisely lunar Quartodeciman celebration, especially characteristic of Jewish Christians.[75] But a much greater part than has been realized up to now was played by solar Quartodecimans: in the context of a solar calendar, they interpreted 14/15 Nisan, the day of the full moon in the Jewish spring month, as the 14th/15th day of their local spring month, or of a fictitious spring month calculated on the basis of the beginning of spring; or, motivated by a fascination for solar symbolism acquired from some source or other, they interpreted that date directly as the day of spring, Julian 25 March.[76] Stern has demonstrated the presence of this already Hellenistic Jewish solar calendar, quite independently of the Books of Jubilees and Qumrān.[77]

In the second and third centuries, the reach of the solar Quartodeciman calculation may have extended to Gaul, Spain, Northern Italy, North Africa, and parts of Asia Minor.[78] Fixed dates for Easter are reflected in fixed birth dates, which, on the basis of variously attested speculation on the equivalence of the date of Christ's death and that of his conception, were set, analogously in each case, on 25 December and 6 January.[79] The widespread currency of these dates – and, if one will, the lack of movable alternatives – demonstrates the firm establishment of the fixed, solar Quartodeciman calculation, and the strong pressure to conform to local, in this case solar, calendars. It appears all the more remarkable that a similar practice did not become established in the case of the Easter Festival.

The main problem may have resided in the early universality of the Easter Festival, a circumstance that did not apply to the birth festival, which became established much later, and in a locally more sporadic manner.[80] Early on, speculative

73 STROBEL 1977: 354, 357.

74 The date is not new: IDELER 1826: 421; MAYR 1955: 324–7. For alternative dates MOSSHAMMER 2008: 50.

75 STROBEL 1977: 357–68.

76 *Ibid.* 368–9, 126.

77 STERN 2001.

78 STROBEL 1977: 370–1.

79 *Ibid.* 155–9; see also TALLEY 1982: 31–8.

80 BALDOVIN 1987: 157–8; FÖRSTER 2000, 2007.

pressure arose to favour the lunar Quartodeciman calculation appropriate to the locality of the original event: this line of thought was eventually to be dominated by the superior astronomical competence of Egyptian Alexandria.[81] A further factor from the earliest times was the historical tradition of the dual dates 14 and 16 Nisan as the dates of Christ's death and resurrection. These could be interpreted not only as those actual dates, but also as 14 Nisan and the following Sunday, although a more strictly historical blossoming of the commemorative festival itself did not occur until later on, and eventually on a massive scale under the influence of the Jerusalem liturgy. In favour of the rapidly popular Sunday was not only the anti-Jewish motif, but also the religious/sociological grounds that 'adjustment to a fixed weekly rhythm represented a considerable relief to the celebrant community, which had already become accustomed to the liturgical significance of Sunday'.[82] This variant applied not only to the lunar Quartodecimans, but also to those of the solar Quartodeciman persuasion; Gaul, for example, at the end of the second century, appears to have celebrated Easter on the Sunday after the day of spring.[83] In the long term, of course, this practice undermined the advantage of a fixed date, which was unavailable to lunar Quartodeciman calculation within a solar reference calendar.

To summarize, we may conclude that this analysis provides no resolution to our initial question in respect of Rome itself. In Rome too, the controversy over the Easter Festival reveals a juxtaposition of different practices, and at the same time a concern for unity that was as powerful at the local level as it was in the Church at large.[84] The insistence on Sunday appears to have represented an early initiative for compromise in this direction.[85] But the diversity that it was designed to counter shows that 'translation' into the standard calendar represented the norm. This indicates, at least indirectly, the normalizing function of the reference calendar: there may also have been little scope for the use of alternative calendars in Rome. That the freedom with which this reference calendar was used was all the greater is demonstrated by the Sunday phenomenon, and the seven-day cult cycle.

10.5 Weekly Cycles

The problem of the week had existed in the Roman calendar ever since, under the decemviral reform, a continuous rhythm of *nundinae* had emerged to join the fixed monthly cycle of orientation days. It was this problem of coordinating the two

[81] STROBEL 1977: 455–6; for all later periods see JONES 1943: 1–122.
[82] STROBEL 1977: 355–6.
[83] STROBEL 1977: 371.
[84] Cf. MAUR 1983: 66; BALDOVIN 1987: 145–6.
[85] STROBEL 1977: 374–88; TALLEY 1982: 26–8: hardly apostolic.

systems that had required the calendar to be put into written form; its nundinal letters could then be used as an aid, but, on a primarily practical level, the problem persisted. When and in what areas the new *nundinae* were able to establish themselves in face of the ancient market days, now orientation days, remains largely a mystery. Cato the Elder, in the middle of the second century BCE, attests to the observance of orientation days in the domestic cult,[86] but the Late Republican discussion of unlucky coincidences of days in the *nundinae* suggests that the nundinal rhythm played an important role. In Plautus, we are told of a *cocus nundinalis*, a cook engaged specifically for the *nundinae*, which suggests a particularly lavish cuisine on those days; but a *lex Iulia*, an Augustan law against luxury, contains no mention of *nundinae* among the three classes of day on which different upper levels of expense applied in respect of meals: 'working days' (200 sesterces); weddings and 'suppers' (1,000 sesterces); but in one breath Kalends, Ides, Nones, and festivals (300 sesterces).[87] The *nundinae* were school holidays; men shaved their beards, and particularly lavish meals were prepared.[88] Later testimony of the special observance of orientation days is lacking.

Although the evidence appears at first sight confusing, tendencies may be discerned. Only in respect of the fixed weekly cycle of *nundinae* can one speak of public, institutionalized 'feast days'. This fits with the account by Macrobius, writing in the fifth century CE, that, in earlier times, candidates for political office had used the *nundinae* as a natural forum for setting out their case.[89] On the other hand, the rhythm of the orientation days tended to be reserved for the continuingly private activities of the domestic cult.

Upon closer examination, it becomes evident that talk of a rhythm is even excessive. The Kalends and the Ides are regularly most prominent, followed by the Nones and other festivals. The many epigrams of Martial that revolve around invitations and celebrations confirm this impression: 'Only rarely is the short toga worn on the Kalends and the Ides';[90] characterizing the quintessence of modest living. It is, again, striking that, just as the birthday of Martial himself was on 1 March, his friends celebrated theirs too on the Kalends.[91] If someone wants to have a birthday on nearly all the Kalends of the year,[92] this highlights the conventional character of the day.[93] A fixed date came into its own precisely for such recurring celebrations.[94]

[86] Cato, *Agr. Orig.* 143; Tib. 1.3.33–4; Hor. *Carm.* 3.23.
[87] Gell. 2.24.14.
[88] Varro, *Sat. Men.* 279; 186; Sen. *Ep.* 86.12; cf. Plin. *HN* 28.28; cuisine: Plaut. *Aul.* 324–5 and Festus, *Gloss. Lat.* 176.27–32 L; Ath. 6.274c (*lex Fannia*).
[89] Macrob. *Sat.* 1.16.35.
[90] Mart. 4.66.3.
[91] Mart. 10.24; 10.29 etc.; 10.87.1; 9.52.2.
[92] Mart. 8.64.2–4: *Uno nasceris octies in anno / et solas, puto, tresve quattuorve / non natalicias habes kalendas.*
[93] Mart. 10.27 and 11.65.
[94] LUCAS 1938. Thus Hor. *Carm.* 3.8.1 and 4.11.14–15 are probably conventional dates. See also *Testamentum Galli cuiusdam civis Romani* (*FIR Bruns*[7] p. 311, II,16–17).

Here we see a picture of a complex society, in which rhythms, whether individually realized or public, combined with the mere leisure to pursue private activities, complemented one another. Political canvassing on *nundinae* was foregone because a sufficient audience was also available on the days in between.[95]

As early as the first century BCE, the eight-day week was joined in Rome by a week of seven days, originating in Babylonia. It probably arose from a conventional subdivision of the month, not from a rationalization of the four phases of the moon,[96] and in this followed a structure similar to that of the Roman orientation days, although the intervals in their case were one day longer. The development of a serial week appears again to represent a subsequent step undertaken as a consequence of the Jewish exile situation, and the encounter with the Babylonian system,[97] although it perhaps did not take full effect until the Hellenistic Period; in any event, these were local reactions whose implementation in Mediterranean Judaism was only limited, and slow to become established.[98]

The Babylonian week reached Rome by two routes, one of them in the form of the Jewish Sabbath observed as a rest day, and the other via the Greek planetary week. The latter again probably did not emerge until the Hellenistic Period, and represented an exclusively astrological concept. The seven 'planetary gods' by order of distance – Saturnus, Jupiter, Mars, Sol, Venus, Mercurius, and Luna – were assigned serially to the twenty-four hours of the day, and then to the one hundred and sixty-eight hours of the week, the god assigned to the first hour of each day also becoming the god of that particular day.[99]

Both the content and the reception of the two forms suggest that it was the planetary week that assumed the central role: an astrological model that combined the simplest divisions of time with a high degree of separation of content, and theologically plausible motivation. It seems as if the Jewish Sabbath, which is scarcely likely to have aroused religious interest, was interpreted from the outset in Rome in terms of the *dies Saturni*.[100] It should be added that Augustan legislation documents a regard for Jewish observance: if the distribution of grain fell on a Sabbath, Jews were allowed to receive their portion on the following day.[101] This does not necessarily mean, however, that the noting of a seven-day week in the Fasti Sabini should be understood as anything other than a reference to an astrological time-structure.[102]

[95] Macrob. *Sat.* 1.16.35; for the public presence see FRASER 1987: 314: 'The *greying of the calendar* is the smoothing out of differences between day and night, the disappearance of distinction among the days of the week and among the seasons of the year.'

[96] WEBSTER 1916: 227–30; LEWY AND LEWY 1943: 1–108, especially 6–25.

[97] MEINHOLD 1905, 1909, 1909a, 1930; WEBSTER 1916: 253–5; LEWY AND LEWY 1943; LASKER AND LASKER 1984 and GOLDSTEIN AND COOPER 1990: 20–1. See also LANDSBERGER 1915.

[98] See BEN-DOV 2008 for the Hellenistic development of calendar-related speculations and experiments. I am grateful for the personal discussion at Göttingen in late 2009.

[99] Dio Cass. 37.18.3–4 and 19: HARE 1832: 4; COLSON 1926: 43–50; GANDZ 1949: 215–24.

[100] Tib. 1.3.17–18.

[101] Philo *Leg.* 158.

[102] SCHÜRER 1905: 26.

The practical consequences of the week's widespread adoption can only be surmised. As far as we can tell, the 'hard' Sabbath practice, as recognized in a succession of legal privileges,[103] and capable of being interpreted by a Roman after the model of the *feriae privatae*, which freed individuals from public duties, was not widely imitated.[104] In contrast, the popularity of the planetary week is reflected in numerous documents, from graffiti to *parapegmata*, and monumental installations such as the basins of the Baths of Trajan, named after the planetary gods, and the Septizonium of Septimius Severus, dedicated in 203 CE, a multi-storey facade with a programme of statues in seven niches.[105] Whether this led to perceptible rhythms in the lives of a substantial number of people must remain an open question: we have no positive testimonies, which fits with the frequent practice in such divinatory systems, to allow the overdetermination of an individual's activities in relation to time to be manifested only in exceptional situations, such as house-building, beginning a journey, and marriage.

Clear evidence for the wide currency of the terminology of the planetary week is provided by Christian inscriptions, especially from the fourth century, which cling to the notion despite the disapproval of the Church hierarchy and theologians, and do not make use of the Christian nomenclature of *feria secunda*, *feria tertia*, *feria quarta* and so on for the days, starting with Monday, a system formulated by analogy with the Jewish Sabbath practice.[106] The term *feria* can be explained only if we acknowledge its theological implications: under the newly instituted rule of Christ every day became a 'holiday'. While this accords with the general tendency in the Imperial Period towards frequent if not daily cult practice,[107] it also points to the explicitly Christian origin of the concept, which found no great resonance among the mass of Roman Christians at the time in question.[108] Generally speaking, we must keep in mind that, at this time, the institutional presence and economic function of the days of the week was not comparable with today's practice, whereby even the average individual, in a highly complex personal field of activity, uses them to manage a great number of autonomous weekly rhythms, but had more in common with the present-day role of zodiacal signs.[109]

The weekly rhythm acquired social relevance where it became the ground-rhythm of a process of religious socialization. This applied first of all to the Jews, but then also to the *Christiani* who sprang from their ranks. The religious differentiation of this group was very quickly reflected at the level of the calendar, not least in the shift of the day of assembly from the seventh day of the week, the Sabbath, to the first day, Sunday. In terms of the day's content, its central importance

[103] GOLDENBERG 1979: 415–22.
[104] GOLDENBERG 1979; Hor. *Sat.* 1.9.69–70; Suet. *Tib.* 32.2; Pers. 5.179–84.
[105] *Hist. Aug. Alex.* 25.5; RICHARDSON 1992: 350.
[106] PIETRI 1984; cf. WORP 1991: 226.
[107] NILSSON 1945; DIHLE 1980.
[108] PIETRI 1984: 74.
[109] Tert. *Ieiun.* 2 p. 275.26–7.

lay no longer in its indivisible essence, but in the shared cult during the hours of leisure at the margins of the day, in the morning and the evening;[110] thus, until Constantine's Sunday legislation, no symbolic, identity-creating function, or soteriological exaltation, pertained to the day, as was the case with the Sabbath.[111]

Relevant to the choice of day was not only mere rejection of the Sabbath, which was, initially, still celebrated in common, but the association of the first day of the week with the Resurrection. The ubiquity of Sunday indicates that it had its origins among the Jewish Christians in Palestine, but it is not possible to demonstrate a direct continuity with the Easter events.[112] In the course of the process, the simple choice of a new day was joined by the consciously polemic refashioning of the Sabbath by deviant practices such as fasting, while rest from work was expressly not associated with the *dies dominicus*.[113]

A law enacted by Constantine in 321 was of decisive importance for the further evolution of Sunday. It defined the *dies solis* as a day free of legal business, and prohibited a whole collection of other commercial activities, the only explicit exception being agriculture.

> Emperor Constantine Augustus to Helpidius.
>
> Just as it appears to Us most unseemly that the Day of the Sun (Sunday), which is celebrated on account of its own veneration, should be occupied with legal altercations and with noxious controversies of the litigation of contending parties, so it is pleasant and fitting that those acts which are especially desired shall be accomplished on that day. (1) Therefore all men shall have the right to emancipate and to manumit on this festive day, and the legal formalities thereof are not forbidden.
>
> Posted on the fifth day before the nones of July at Cagliari in the year of the second consulship of Crispus and Constantine Caesar. – July 3, 321.[114]

This corresponds to the ruling in the *Codex Iustinianus*:

> *Omnes iudices urbanaeque plebes et artium officia cunctarum venerabili die solis quiescant. ruri tamen positi agrorum culturae libere licenterque inserviant, quoniam frequenter evenit, ut non alio aptius die frumenta sulcis aut vineae scrobibus commendentur, ne occasione momenti pereat commoditas caelesti provisione concessa.*[115]

[110] RORDORF 1968: 215–62, 1981: 152–5; BAUCKHAM 1982: 237–9.

[111] KLINGHARDT 1991: 218, 230.

[112] BAUCKHAM 1982: 234–6.

[113] MEINHOLD 1909: 65–76; RORDORF 1968: 80–8, 104–18; KLINGHARDT 1988: 244–52; cf. BACCHIOCCHI 1977: 165–235

[114] *Cod. Theod.* 2.8.1: *Imp. Constant(inus) A. Helpidio. Sicut indignissimum videbatur diem solis veneratione sui celebrem altercantibus iurgiis et noxiis partium contentionibus occupari, ita gratum ac iucundum est eo die quae sunt maxime votiva conpleri. atque ideo emancipandi et manumittendi die festo cuncti licentiam habeant et super his rebus acta non prohibeantur. P(ro)p(osita) V Non. Iul. Caralis Crispo II et Constantino II CC. Conss.* Tr. by Clyde Pharr, *The Theodosian Code*, Princeton University Press, 1952.

[115] *Cod. Iust.* 3.12.2. DÖLGER 1940: 229–30, 235; RORDORF 1968: 162ff.

(All judges and city inhabitants, and the workshops of all artisans, should rest on the venerable Day of the Sun. Those in the countryside should tend to the cultivation of the fields legally at their discretion, for it happens frequently that no other day is more apt to entrust the plants to the furrows or planting-holes in the vineyards, in order not to lose the advantage of the right moment granted by heaven.) (Translation by Rüpke, 2010)

An examination of the text as it has come down to us in the Codex Iustinianus in particular reveals that the concept is new. *Venerabilis* and *quiescere* have no status as *termini technici* under secular or sacral law. Judges comprise the first category included in the area of validity of the law: this corresponds with the traditional notion of normative authority.[116] The extension of the scope of the day of rest to the urban population and workshops, on the other hand, appears to be quite new; that what is meant here is actual rest from work is shown by the exception granted to agriculture, and the subsequent comprehensive justification of that exception in the text.

Although the ruling is delivered in a religiously neutral form, there can be no doubt that its motivation is Christian. In the law's wording, *venerabilis* refers to a cult practice that must comprehend more than an astrologically favourable disposition on the 'day of the sun'.[117] This interpretation coincides with that of Eusebius, who, in his biography of Constantine, includes the law in a series of other rulings in the area of court life and the military, which ultimately encompassed the explicit institution of Sunday as a day devoted to the service of God.[118]

The association of Sunday with rest from work also constituted an innovation. Constantine could nevertheless count on the agreement of the Christian Church, which was thus accorded a protected status that was 'set in stone', and largely secure against competition. The service of God on Sundays was no longer assigned to the extreme margins of the day; and this was a by no means inconsiderable advantage, when one considers that, already two decades earlier, a council in Spanish Elvira had enacted a 'Sunday observance' statute that, admittedly, was very liberal in comparison with the scholasticism of Thomas Aquinas: whoever, as a resident of a town (here the problem of control plays a part), fails to attend church service three Sundays in succession, shall be excluded for a brief period in order to expose and punish such behaviour.[119]

[116] See *Cod. Iust.* 3.12.1 of 305 BCE.

[117] Cf. *Cod. Theod.* 2.8.18: *Solis die, quem dominicum rite dixere maiores, omnium omnino litium, negotiorum, conventionum quiescat intentio; debitum publicum privatumque nullus efflagitet; nec apud ipsos quidem arbitros) vel iudiciis flagitatos vel sponte delectos ulla sit agnitio iurgiorum. (et) non modo notabilis, verum etiam sacrilegus iudicetur, qui a sanctae religionis instinctu rituve deflexerit.* See also DÖLGER 1940: 236–7.

[118] Euseb. *Vit. Const.* 4.18–20; especially 4.18.2; Sozom. *Hist. eccl.* 1.8.11–12.

[119] *Concilium Illiberitanum* (Synod of Elvira) can. 21 (*PL* 84.304; RORDORF 1972: no. 110).

In the longer term, this designation of Sunday as a day of rest facilitated the adoption of Jewish *theologoumena* for the Christian *dies dominicus*. However, the extent to which the Sabbath served as a model for Constantine's ruling remains questionable. We might gain a sharper focus on the law's intention and models by not confining our historical perspective to the Sabbath, Sunday, and the planetary week: the old Roman concept of *feriae* certainly comprised the element of rest from work, or at least a restriction of work.[120]

Ancient discussion of the subject, generally speaking, covered only agricultural activities. A ruling that involved not only the highest-ranking priests, the *flamines maiores* and the *rex sacrorum*, but also the *pontifices*,[121] suggests that an analogous reference to urban activities may have been possible, and, setting aside for a moment the question of realization, was actually drawn: the norms generally applying to *feriae* involved the above-named officials to a particular degree. Such rules covered not only their own labour, but also the *opus* of others. The test of contemporaneity was visual contact, and it was the sight of work that was incompatible; the problem of infringements against the *feriae* rules was anticipated by means of a 'time-bubble' accompanying the official in question: a herald ran ahead, and compelled those along the way to interrupt their work on pain of punishment.[122]

A reference to the concept of *feriae* according to sacral law is also indicated by the specific treatment of agricultural activity in Constantine's statute. Although the injunction to rest is explicitly addressed only to the urban population, and rural labour left generally unregulated, the grounds given for this dispensation make it clear that the exception thereby established has in principle to be justified. The grounds given, however, are beset by so many contingent factors that no consistent ruling is actually offered, and those affected are left to decide for themselves, free of legal pressure. The activities listed as examples of exceptions – the sowing of a field already ploughed, the planting of vines in furrows already prepared – respect the principle that on *feriae* the ground must not be worked with iron (ploughshare or spade).[123] The indication that the Sunday in question is an ideal occasion, an opportunity offered by God's providence, which might come to nothing if let go by, corresponds to the principle expounded by the *pontifex maximus* Q. Mucius Scaevola, that those things are allowed on *feriae* that might suffer by being neglected.[124]

This was the position regarding Sunday regulation until the sixth century, when the council of Orléans (538) established as a norm the ban on agricultural activ-

[120] Sen. *Dial.* 9.17; DÖLGER 1940: 235.

[121] Serv. *auct. Georg.* 1.268; see also Plut. *Quaest. Rom.* 25.

[122] Macrob. *Sat.* 1.16.9; Festus, *Gloss. Lat.* 292.3–7 L; Paul. Fest. 250.15–17 L; 293.1–3 L.

[123] Serv. *auct. Georg.* 1.268.

[124] Macrob. *Sat.* 1.16.11; Serv. *Georg.* 1.270; Pallad. 1.6.7.

[125] *Concilium Aurelianense* (Council of Orléans) can. 31 (RORDORF 1972: no. 139); cf. Verg. G. 1.270.

ity.[125] Although repeatedly raised down the years, the injunction proved unrealistic, and remained broadly ineffective into the Modern Period.[126]

Our presentation of the background of the 321 ruling in respect of sacral law takes away nothing from its innovative character. In terms of the casuistry of the *feriae* debate, the religious tone of *commoditas caelesti provisione concessa* is new; the main novelty, however, is the fact that these religious requirements were made the subject of state regulation at all. In the history of law and religion, this aspect does not make the statute unique; but it confirms its place in a line of development that is noticeable especially from the third century onwards: religious institutions, and not merely undesirable foreign cults, are increasingly made the subject of state legislation. In this way, the process that, in the Imperial-Period calendar, had changed the state *fasti* into a direct medium of religious regulation by recording *feriae, quod* instead of simply *NP*, was continued on a new level.

Although it is impossible to overestimate the effect of such a far-reaching ruling, the mere setting of a norm did nothing in itself to create mechanisms of control or sanction. The constant repetition of many laws, which is still discernible even in the great digests, at the same time reveals their ineffectiveness.[127]

Eusebius' account embeds the Sunday law – the still more comprehensive formulation of Friday observances, mentioned in Eus. *Vit. Const.* 4.18.2, has left no further legislative traces – in measures that reveal where Constantine's opportunities for implementation lay: in his own palace, and within the overall regulative structure of the army. Even here, there is no talk of a generalized rest from work or interruption of service: rather, Sunday is a day for worship in common, in the form of a prayer; Christian soldiers are additionally relieved from duty for the period of the religious service.[128] These regulations also show the extent to which Constantine's thinking followed traditional paths: on the one hand, so little note was taken by the military of interruptions of service as a consequence of *feriae publicae*, that trials concerning military discipline were not even affected by those days;[129] on the other hand, when a man was called up, even the military respected delays due to the observance of indispensable cult obligations and *feriae familiares*:[130] not a precise analogy, but nevertheless an available element of respect for an individual's religious obligations.

The reliance on existing institutions is also demonstrated in a permission by Constantine for Pannonia, attested in an inscription, to hold a market – *nundinae* – on the *dies Solis*:[131] the functional equivalence between Sunday and

[126] Concilium Matisconense II (Council of Macon II) can. 1 (585 CE); Concilium Narbonense (Council of Narbonne) can. 4 (589); Gregory of Tours, *Hist. Franc.* 10.30; *Iul.* 11; *Mart.* 3.3. 10.30).

[127] Demandt 1989: 254.

[128] Euseb. *Vit. Const.* 4.18–20.

[129] Ulp. *Dig.* 2.12.9.

[130] Cinc. *Gell.* 16.4.4–5; Rüpke 1990a: 69–70.

[131] *CIL* 3.4121.

nundinae, which would subsequently make the latter superfluous, can be seen clearly here.[132]

As regards the law's reception, it seems to me important to observe that further instances of Sunday regulation concentrate only on the juridical element of the court holidays, handed down solely in the *Codex Theodosianus*. Like today's British bank holidays, they functioned as a '*leitmotiv*' of the holiday structure; they alone were the subject of state regulation.

The judicial system alone served as defining principle for other holidays, and for their status. The total ban on activities lapsed for the longer *feriae* periods of the harvest, Lent, and also Easter; here, detailed rules differentiated the areas of legal activity that were irreconcilable with the character of the period in question from those that were appropriate:[133] a procedure that may be seen as analogous to the distinction between *NP* days and *N* periods. In any event, the mere existence of these rules shows that the Republican differentiation of the legal system, also politically based, between *in iure* – to which the *fasti* applied – and *in iudicio* – on which they had no influence – no longer played a central role.[134]

Finally, laws were passed to address a problem that persists to this day, even though it is no longer defined in moral terms but in terms of labour law. Designation of Sunday as a day of rest creates leisure-time, which demands appropriate services, and thus the work of others. This aspect is still relevant today; leisure not only produces problems of synchronization and social dislocation in complex societies, but – and this was of paramount importance to ancient legislators – it gives rise to activities of a distinctly unholy character: repeated attempts were made to prohibit *obscenae voluptates*, and games of all kinds.[135]

Whatever the level of implementation we might imagine to have accompanied these regulations (and normative texts are always bad sources in such questions), one consequence of Constantine's law may be regarded as certain: after 321, if the *fasti* were to maintain their utility to their owners, of whatever religious provenance, as a representation of the valid calendar of record, they had to accommodate the letters of the seven-day week. The Fasti Filocali reflect this requirement, but, with their continued use of the nundinal letters, they show that no rapid process of replacement had taken place. One hundred and twenty years later, the Fasti Polemii Silvii lack both features: a further indication that the later Christian *calendariae* with their Sunday letters do not represent an unbroken continuity with the *fasti* tradition, for all their similarity.

Although the calendar entry for Sunday was to be implemented in much the same way as the entry for the *nundinae*, in that the corresponding information had to be conveyed orally, the associated loss of precision was more serious in the

[132] A law of 389 (*Cod. Iust.* 3.12.6; see below) removed the *nundinae* as a legally relevant instititution (see pr. and § 4; UNGER 1895: 514). The term is found in the *Codex Theodosianus* only in the sense of 'market'.

[133] *Cod. Iust.* 3.12.3–9, e.g. 3.12.7 of 392.

[134] In questions of detail, it was still important in the second century CE: Ulp. *Dig.* 2.12.2.

[135] *Cod. Iust.* 3.12.6 (389); *Cod. Theod.* 2.8.20 (392); *Cod. Iust.* 3.12.9.2 (469) etc.

Christian case: Sunday was a vital religious date, to a much greater extent than the market day, and, if the calendar was to be used symbolically, in a way comparable to its use in the Augustan period, it required more by way of coding than an uninterrupted sequence of letters from *A* to *G*.

10.6 *Fasti Christiani?*

The representation of Sundays in the calendar is highly significant insofar as, after the celebration of Easter had been linked to the Sunday, the most important Christian festivals – and more still were considered! – were anchored in the structure of the seven-day celebration of the *dies dominicus*. An early member of this group was the Whitsun festival of Pentecost (*Pentecoste*), 'fifty days', that is to say seven weeks, after Easter.[136] But the process continued. The period immediately preparative to Easter was gradually extended to include the *Quadragesima*, the period of Lent, which eventually comprised six Sundays; this move was pursued with particular intensity in Rome, where one by-product was the development of the stational liturgy.[137] The further evolution of this more densely structured core period occurred at a later time (Trinity Sunday and Corpus Christi, for example, in the thirteenth and fourteenth centuries):[138] no new creations emerged, however; the early coherence of the entire annual Sunday cycle was merely underlined. This coherence outside the festival periods was created during the first centuries by the continuous, though selective, reading of entire books of the Bible (cycle of readings);[139] the evolution of Sunday liturgies into a multitude of isolated, freestanding thematic celebrations did not yet play a role at this period.

The recorded dichotomy between the Julian calendar and the Christian calendar of festivals tells only half the truth. Local calendars, here in particular the Roman calendar, displayed considerable persistence in face of the whole concept of a weekly rhythm and a relativist calculation of Easter; thorough analysis of the tradition clearly shows that for a long time Easter itself was more closely linked to fixed (solar) dates than was discernible by the end of the process. However, clearer testimony for the significance of local calculations of time is provided by the *sanctorale*, the Calendar of Saints.

The veneration of saints as advocates, or intercessors, developed during the course of the third century, in part from older, Jewish roots, and from the early custom of celebrating 'blood witnesses', those fellow Christians for whom the promise of resurrection had already been realized. This attempt to transform

136 TALLEY 1982.

137 KIRSCH 1926; WILLIS 1968; BALDOVIN 1987: 143–66.

138 MAUR 1983: 195–206.

139 See MAUR 1983: 44.

the omnipotent but therefore also inaccessibly distant godhead into a tangible opposite number capable of being addressed occurred in a world that appeared increasingly to defy political structure, and was amenable only to the control of powerful individuals: a social model based on patronage. This problematic situation explains the doubly local reference of the martyr cult,[140] which primarily relates to a local personality, and is exercised at a location that promises closest possible proximity to the venerated individual: his (or her) grave.[141] In the context of this investigation, we need to give attention only to that aspect of the phenomenon that saw the institutionalization of the spontaneous cult, and the perpetuation of the anniversary day, whose celebration was customary in the individual cult of the dead,[142] in calendar form.[143]

Such expression is first found in the form of a list of *dies depositionis* in the *Chronograph* of 354.[144] Three points are of particular importance in defining the character of this list: the addition of details of locations indicates a practical cult; although the list of names particularly mentions local martyrs,[145] the location *Africae*, given for one of the non-Romans, gives rise to difficulties, as it leaves open the question of a cult practice within the city of Rome; the Christmas festival, *nativitas*, heading the list, detracts from its specialized character, and points in the direction of a liturgical calendar. That this fixed date is linked with martyrs' anniversaries, and not with the Easter cycle where it conceptually belongs, appears technically plausible: it was not until the Early Middle Ages that 25 December won a position in the weekly structure associated with Easter, through the creation of a preparatory period by the inclusion of a series of Sundays (Advent).[146]

The absolute, Julian dating of these memorial days, although self-evident, is still worthy of note. In principle, inclusion of the *Sanctorale* in the Easter cycle, the *Temporale*, would have been conceivable; the resulting dates would have remained stable relative to Easter. The fact that this did not happen confirms once again the primacy of the local calendar, which sanctioned use of the appropriate 'foreign' calendar only in the case of the date of Easter, based as it was on a geographically remote event, and for those other dates directly dependent on it.

The status of the list and its provenance remain unclear, simple though its form may be. At the time it was copied, some two decades before being bound into Filocalus' collected works,[147] it may already have constituted a selection.[148] It is possible to understand the essential character of the individual elements of this and later lists. Most of them are dates of burials, *dies depositionis*, although these

140 BALDOVIN 1991: 39.

141 KLAUSER 1960: 30; P. BROWN 1981: 1–22; DINZELBACHER 1990; J. MARTIN 1987; 1988.

142 FÉVRIER 1977; SCHEID 1993.

143 KIRSCH 1918: 208–13.

144 MOMMSEN 1892: 71–2.

145 FRERE 1930: 8; DENIS-BOULET 1960: 55; MAUR 1994: 137.

146 TALLEY 1981: 117–20.

147 See SALZMAN 1990: 43–4.

148 KIRSCH 1924: 238–9.

may also be described as birthdays, *dies natalis*. This concept should not be overburdened with theological significance.[149] The practice of celebrating days of death as 'birthdays' is a phenomenon that goes back to the Early Hellenistic period, although there it does embody a clear theological message: the day of Alexander's death is the birthday of the new god, and is celebrated as γενέσια.[150] But the combination of dates is frequently found without this massive implication; the attempt is often made in Late Antiquity to move the *dies depositionis*, which is the date distinguished by the cult, and not simply identical to the date of death, closer to the actual *dies natalis*.[151] Then again, we find dates that relate to secondary cult events, such as the construction of a church over the grave.[152]

The purpose of such collections defeats simple analysis. At later periods, we find comparable lists forming part of *sacramentaria*, or of other texts designed to be used in the liturgy, and tailored to the needs of a particular priestly order (*ordo*).[153] In the case of Rome, the episcopal collections are of vital importance, the earliest of these being contained in the *Liber sacramentorum anni circuli* of Gregory the Great: this was a stational *sacramentarium* belonging to the bishop of Rome;[154] although highly influential, owing to its restricted function it had to be modified repeatedly.[155] Similar is the somewhat more recent *Gelasianum*, which is probably based on Roman sources.[156] The *Martyrologium Hieronymianum*, from the beginning of the fifth century, should also be assigned to this central group; its structure is similar to that of Filocalus' list,[157] and it comes from a period of critical importance for the further development of calendar-like liturgical *corpora* in the West.[158]

There is a possible case for thinking of the compilation in the Chronograph of 354 as the prototype for an episcopal *feriale*. The complex structure of the Roman community suggests the existence of a literate central bureaucracy, not inferior to that maintained by comparable traditional Roman priesthoods. Structured on an annual basis, the list provides the portrait of a cult community founded purely on a 'stational' practice, systematically visiting all cult locations,[159] and therefore provided with dates and organized according to those dates. The structure for the commemoration of the dead found in place by the Christians, and the fact that it was frequently secured by foundations and colleges, suggests that, in Rome,

149 See DELEHAYE 1933: 35–6; Tert. *De coron.* 3.2; *De monog.* 10.4; *De resurr.* 13.2 (see MAUR 1994: 92–4).
150 HERZ 1978: 1142.
151 DUVAL 1982: 571.
152 KIRSCH 1924: 239, 241; DIX 1945: 377ff.
153 DIX 1945: 365.
154 GAMBER 1956: 8; MAUR 1994: 141–4.
155 MAUR 1994: 9.
156 VOGEL 1981: 67 f.
157 KIRSCH 1924; LIETZMANN 1927: 29–82.
158 GAMBER 1970: 91. 99–106.
159 BALDOVIN 1987: 145–6; KIRSCH 1926: 5.

the administration of burial grounds (*coemeteria*) begun by the Church during the first half of the third century formed the nexus of a memorial structure, also officially organized, and compelled by pressure of circumstance to focus primarily on martyrs and bishops.[160]

The question arises as to whether the organization of the festal calendar can be categorized as a primarily pastoral measure.[161] It did not become possible in the form of a written calendar until much later, when the appropriate technical means of reproduction became available. The unified audience implied by the term 'pastoral' for a *feriale* of the entire city of Rome, i.e. of Rome as one community, is not conceivable. It is impossible to imagine Rome's Christian population, probably already segmented into twenty titular churches (*tituli*) – not to speak of heterodox groups – as the subject of such a detailed central festal list. The degree of liturgical integration represented by the stational liturgy and the *fermentum* (the distribution of consecrated pieces of bread) remained, in practice, at a very elevated level in the hierarchy; it did not extend further until the sixth century.[162] Thus, even taking Rome's festal tradition into account, we are left with the much simpler hypothesis that the lists of memorial days primarily represented a means of managing the claims of the dead to cult memorialization.

Truly pastoral objectives at the level of cult organization, which is fundamental to the list, do become clear in the placement of memorial days on traditional feast days, and the avoidance of the Easter period.[163] The same does not apply to Filocalus' list as such. The interests of the simple community member might much rather be reflected by a topographically structured itinerary, such as has been revealed to constitute the earliest stage of the *Calendario Italico* of 425/430.[164] This is the leading level of activity, where we should also locate the supra-regional exchange of saints' festivals, a process by which, in the course of Church history, phases of extreme local differentiation repeatedly alternated with phases of strengthened, centralizing bids for uniformity, partly in the form of the imposition of the local Roman calendar.

Independently of local liturgical practice, the *feriale* of saints (*Sanctorale*) soon combined with the religious *tabula rasa* of the Julian calendar; such a link was unproblematic for this type of date, and would, moreover, be lasting and profound; it would find expression in the widespread use of saints' days in the Middle Ages for dating purposes. The calendar's liturgical and pragmatic functions were quickly to be joined by an independent process comparable to the development of Fulvius' *fasti*: the lists of saints are interpreted historically; the calendar is filled with relevant information such as additional names, and historical settings in place

160 STUIBER 1960 with reference to *CIL* 14.326; cf. P. BROWN 1981: 25–49; FONTAINE 1982: 22–5.

161 Thus PIETRI 1983: 120.

162 BALDOVIN 1987.

163 PIETRI 1984a: 301–2.

164 LUCCHESI 1978: 147–9. Cf. FRERE 1930: 9–10.

of cult localities, giving it popular appeal, both as a historical document and as a theological-cum-hagiographic compendium.[165]

In the Carolingian-Ottonian period, compilations of this type developed into versified calendars, to become particularly popular in the British Isles: only a completed list of saints makes possible a form such as the so-called *Metrical Calendar of Hampson*, now enabled to devote a separate line to each day of the year.[166]

Then there is that form of calendar that, in a solitary example from Naples, possibly originating in the mid ninth century, offers at the least an indirect continuation of the Early Imperial Period marble *fasti* as mediated by book calendars: a 'volume' consisting of two marble tablets measuring in all five and a half metres wide by 0.76 metres high, and listing the twelve months of the year, separated by semi-sculptural round pillars. Each month is depicted in two columns, and, on the basis of a caption of the type *M(ensis) Februarius d(ies) XXVIII*, lists all the dates of the month in ascending, purely numeric order, providing details of the saints: *dies natalis* (*N*), passion dates (*PĀS*), *dies depositionis* (*DEP*) *et al.*, with rare exceptions of days that remain empty (e.g. 8 and 14 March); other festivals sometimes replace those of the saints (Epiphany, Purification of the Virgin). Each day ideally features one and only one saint's name, or, at most, complementary pairs; cumulations of dates are avoided.[167] A caption above the first six months reveals that the inscription's primary intention is to demonstrate the great number of the saints. It is not the year presented in calendar form that takes pride of place: this aspect is not visibly developed anywhere within the monthly double columns. Rather, the schema of the three hundred and sixty-five days embracing all of time serves as a symbol of completeness. If a saint can be named for every element of 'all of time', in the final analysis that multitude of names serves to negate the finiteness of the numbers of the saints.

Even though, in terms of form, a clear line can be traced back to the calendar of Gn. Flavius, the 'filling', and thus the calendar's intention and function, have in the meantime changed radically. While, in the Fasti Filocali, cultural changes were restricted to the calendar's own context, and, in the Fasti Polemii Silvii, they formed a combination in which the tradition of the *fasti* was still clearly recognizable, from this point onwards, superficially at least, the transformation is complete. The earliest extant Christian stone calendar thus does not represent a resumption of the tradition of the Early Imperial Period, but a new beginning. While the technical aspect may still be reminiscent of the 'sub-antique', to anyone whose interest lies in religious function the Middle Ages are already clearly recognizable.

Of course, all these calendars had long since ceased to be *fasti* in any concise sense of the term. Already in 389 we see a law of the emperors Valentinian, Theodosius, and Arcadius, bringing a definitive end to the era of calendars showing

165 HENNIG 1980.

166 MCGURK 1986: 90–111.

167 LECLERCQ 1925; ACHELIS 1929; MALLARDO 1944–46.

dies fasti and *nefasti*, and replacing detailed information of that genre with a *feriale*. This new conceptualization could no longer be integrated even into a 'cleansed' calendar, no longer even lent itself to being visualized in the classical form of a list of dates.[168] With this ruling, coming at the end of a long series of measures of 'persecution against pagans', the *fasti*, along with the Kalends, the Nones, and the Ides, become mere calculating instruments: the unity striven for during the Imperial Period is finally shattered. By this measure, the information for whose sake Flavius published the first *fasti* now largely disappear from them. In this changed situation, four centuries after the end of the Republic, the direct political consequences may have been negligible, the juridical consequences not particularly grave. In public terms, in terms of publication, however, we return to the state that pertained before Flavius: not exactly a full circle, but an appropriate end to the historical process.

[168] *Cod. Iust.* 3.12.6.

11

The Calendar in the Public Realm

Calendars are among the most important instruments used by a society to organize its own affairs in respect of time. This involves firstly a technical dimension: in many cultures, an objective for calendar-related constructs is a certain compatibility with the solar year; that is to say, points in the year that are, according to the socially defined 'calendar', identical should also in some way coincide with equivalent points in the solar year, and thus comparable climatic conditions, over a relatively long period of time. The pure lunar year of the Islamic calendar directed against the religions of pre-Islamic Arabia represents a rare, polemical construct, which produced distinct deviations within a few years.

In Rome, at various times we see a concern to make technical improvements to the calendar. The desire first becomes evident in the reform at the end of the fourth century BCE, and persists without a break in relation to the problem of intercalation certainly from the third century onwards, rising to a new level in the *lex Acilia* at the beginning of the second century, with the transfer of the intercalation process to the *pontifices*. In principle, the Republican calendar depended operationally during this time on the commonly practised expedient of empirically based intercalation using authentic lunar months, which permitted short-term differences of at most one month in the calendar year as against the solar year, but was capable of maintaining long-term parity. It was not until the Julian reform that the Roman calendar departed from this model, and the intercalary period was reduced to the intercalary ('leap') day, still familiar today, interposed every four years without need of separate deliberation.

The significance of the reform at the end of the fourth century BCE in general historical terms lay in its translation of the technical instrument into written form, which provided the means to stabilize the solution once found. While this particular reform did not fix the relationship between the running 'weeks' and the

---

The Roman Calendar from Numa to Constantine: Time, History and the Fasti, First Edition. Jörg Rüpke.

months, it did make it possible to follow that relationship. We know of no other comparably practical instrument in pre-Roman or non-Roman Antiquity; the schema used for the calendar's visual presentation remained in place until the typographic revolution brought the annual reprinting of calendars within the realm of reasonable affordability. It was also the logical consistency of the fixed, written calendar that facilitated the removal of intercalation's remaining degree of uncertainty, which had not been sufficiently alleviated by primitive rules. It was this aspect that Caesar had in mind, with his truly revolutionary simplification and minimization of the period and frequency of intercalation. It was now no longer at all necessary to represent the intercalary period graphically in calendars. The published version of the calendar was more easily readable than ever before.

In the particular form it took, the calendar established a substantial framework for the concrete use of time in a society. The detailed nature of such real strategies in relation to time depends essentially on the degree of diversification reached by the society in question. In principle, as distinct social subsystems emerge, they elaborate their own subordinate calendars, that is to say their own schemata to superimpose on the common structure. This process can be observed in the case of Flavius' reform in Rome in his differentiation of distinct market days (*nundinae*) from the orientation days, which were likewise separated within the month by intervals of 'nine' days, and previously served as 'proto-*nundinae*'. The grounds for the creation of a continuous system of 'weeks' running across the months' limits may have resided primarily in the economic sphere.

Rome's parallel, religiously motivated conservation of the old system was the critical factor that differentiated its model from the serial week that evolved only a little earlier in Mesopotamia and subsequently Palestine; although in these cases that week consisted not of eight days, but seven, a form that, eight hundred years later, in the shape of the Christian Sunday cycle with its precursors in the Jewish Sabbath and the Hellenistic planetary week, would displace the Roman system. Considered in the long term, in Rome's case the price incurred by the pressure for differentiation that would arise from this parallelism was the loss of the weekly rhythm, which had been an uncommonly effective source of religious cohesion.

The differentiation of the religious calendar had one fundamental consequence for the general technical development of Rome's calendar: as the religious calendar was closely associated with the lunar month, and, in its concern to maintain old institutions as far as possible, persisted uninterrupted with a counterfeit style of lunar month, the main calendar of reference, in its quest for renewal, which was itself facilitated by committal to written form, was released from the necessity to have regard for the actual lunar cycle. It thus found a characteristic shape that, while adhering formally to the milestones of the empirical lunar month, in substance, that is to say in its crucial determination of the months' lengths, parted from the lunar month, and became *de facto* a solar calendar. This provided the necessary conditions, in a world full of lunisolar calendars, for the acceptance of Caesar's maximal juxtaposition of the calendar with the solar year.

With the Hellenistic planetary week (which, however, appears to have had little influence on actual weekly rhythms), with the Jewish Sabbath, which applied only to a minority, and with the slow evolution of the Christian Sunday festival, the parallelism became manifold. Eventually, beginning with Constantine's Sunday legislation, the Christian weekly rhythm emerged triumphant. In the coexistence of these systems, we can see in a nutshell the problems involved in the relationship of social minorities with the generality of society. In establishing norms that give a structure to time, in either forbidding or prescribing particular activities, one group in society impacts enormously on other social subsystems. The consequences of such interdependence are tolerable for those other subsystems so long as the effect on the total 'stock' of time is insignificant (the obligation to say a short prayer, or to sacrifice on the day in question), or applies only to a minority (the Jewish ban on Sabbath work). The first condition seems to have applied, alongside norms that were somewhat vague, to the relationship between orientation days and *nundinae*; our sources, at least, complain of no difficulties. Of course, these interdependencies apply in both directions: society's Christian identity was also targeted by the changes to working times expressly occasioned by the French Revolutionary Calendar and the Soviets' five- and six-day weeks.

How should we assess Roman conditions from the point of view of long-term social development? That the differentiation of functionally specialized systems is accompanied by a redefinition of functional hierarchies is illustrated in Rome's case by the dominance of the *nundinae* over the days on which they fell: a dominance that was neutralized by special dispositions only in isolated instances. It is also the case that the separation of agricultural activities from the rituals serving them, brought about by determinations in the calendar, must be considered under the aspect of differentiation, and each ritual in question interpreted accordingly.

A clearer view of the problems associated with the differentiation of calendars for different subgroups within society may be obtained by considering the present-day situation. Dominance by the commercial sector is exemplified by the fact that discussions of 'general' working hours have long since ceased to relate to economic activities overall, and are now confined to isolated sectors. The general limits suggested in the context of controversies concerning shop-opening times and the free weekend have long been overtaken by shift work, the transport sector, and the leisure industries. The disappearance of periods reserved for particular activities, and the importance of 'leisure' activities, defined as non-participation in the market, force old systems to adopt new strategies: diversification of the commercial offer over extended periods of time, shorter 'time slots', the more intensive use of available windows of time. Mass mobility further exacerbates the problem for highly industrialized societies, and traffic problems (rush hour, seasonal peaks, holiday jams) compel the 'staggering' of working hours and holidays. Here too, individual gains in flexibility have a remedial effect, opening up new opportunities for small groups, while large-scale social organizations are forced to turn more intensively to the strategies described ('prime-time slots' in place of open-air demonstrations). This also gives us some idea of the nature of the processes concerned. For the sake of face-to-face contact between participants, increasing problems of coordination, arising from increased functional differentiation, must be compensated by segmentation. In Germany, the district street party of the last years of the twentieth century replaces the national day of the nineteenth.[1] At the level of the social collective, however, the calendar loses its shape, becomes blurred. In place of a society coordinated by the calendar, we have agendas negotiated between individuals or within specific groups.[2]

The keyword 'segmentation' takes us back to the situation in Antiquity. Here, the limited possibilities of communication technology, and of media of communication in the broadest sense, set narrow limits to functional differentiation. The degree to which differentiation through segmentation was a determining factor is often hidden by the concentration of historiography and the monumental record on the functional elite, and on territorially extended but highly precarious systems of rule. In practice, internal cultural acceptance of such segmentation, and its self-evident character, made many problems entailed by functional differentiation less acute: the public realm remained as a rule local.

The choice of the date for Easter illustrates problems involved in the creation of time-related structures. It is striking that the date was not finally settled until Constantine's Sunday legislation had created a stable frame of reference throughout the Empire for the linking of the Easter Festival with Sunday, so that its relationship to the Julian calendar could be restricted to the definition of a target range, the Easter limits. In considering the influence of fundamental calendrical rhythms on the placement of such festivals, we should think again of the case of temple-dedication days. Here we saw an unmistakable orientation towards the religious rhythm of the orientation days, which despite a degree of distribution throughout the year , resulted in a concentration around particular days of the month.

Knowledge of and regard for these fundamental rhythms were not sufficient for the everyday, detailed planning of events. In the early period, the festivals were largely absorbed into the rhythm of the monthly orientation days, in particular through being positioned at the nundinal date between Ides and Kalends; but, already at the beginning of the Republic, the position was to a varying degree more complex.[3] In the political, juridical and religious sectors, matters were much complicated by social developments in the fifth and fourth centuries. This situation must also have found expression in the area of the calendar, in the form of problems of coordination. Particularly affected from the religious point of view were those *feriae* that the calendar would later mark *N(efas) P(iaculum)*, and which cor-

[1] On the latter, MAURER 1991: 122–7, with further references.

[2] FRASER 1987: 314–15.

[3] We need think only of the *dies natales* of the temples founded *c.*500 BCE.

responded chronologically to the *dedicatio* of land, and especially temple-building. Other religious rules were associated directly with the protection of orientation days; they defined the *dies postriduani* as *dies atri* ('black', thus inauspicious days), other days as *dies proeliares* or *comitiales*, and a more indeterminate class as *dies religiosi*.

These many and diverse qualifications affected the public realm, but were not themselves published. The monthly announcement of the *feriae* occurred only orally, and not until the Nones of the month in question. The assembly before the *rex sacrorum* remained the sole authentic mode of publication on the part of religious institutions into the Imperial Period. Beyond this, more complex norms of behaviour in respect of particular days may have been left to entirely traditional modes of knowledge and communication; religious motives may not necessarily always have been the decisive factor in the barring of particular days to people's assemblies and *legis actiones*.

It is difficult for us to judge whether the amount of obstruction and discrimination hinted at in Late Republican sources actually pertained in everyday political and juridical affairs. What is crucial, however, is that, from the perspective of the reforming ideas of Appius Claudius at the end of the fourth century, such practices were classified as anachronistic. The purpose of the codification of the *fasti*, carried out at Claudius' behest by Gnaeus Flavius, was to provide a remedy by paring down the authority exerted by the traditional custodians of knowledge, above all the *pontifices*, over the political and juridical ramifications of the religious qualification of time.

Presenting the various types of information within the confines of a written calendar caused Flavius great problems. The concentrated, tabular form compelled brevity and succinctness, and no doubt also simplifications and summary classifications that had not previously existed: the *dies E(ndoitio) (Exitio) N(efas)* may have sprung from this struggle to reconcile precision and credibility in terms of sacral law with the practicalities of classification and representation.

Of central importance to the further development not only of the Roman calendar, but of calendars in general, is the independent life, already discernible here, acquired by the written version of the calendar as against the calendar as a practical element of everyday life. In the period immediately following, in 287 CE, the *lex Hortensia* had to deal with problems of political and juridical classification that had been thrown up only by the acuity of definition in the terminology of Flavius' *fasti*. This independent life can also be seen in the festival names, in the neologisms developed for the sake of clarity in abbreviation, and in the terms promoted from the colloquial realm to the status of *termini technici*. Classical Philology and the History of Religion have largely neglected the task of understanding the status of these terms, and, in their analysis of Roman religion, of more sharply distinguishing the level of the *fasti* from everyday practice.

But even the message underlying this criticism remains too imprecise. It is not merely that the written form of the *fasti* must be distinguished from 'the' calendar;

a further distinction must rather be drawn between society's calendar of reference and individual, sectional calendars, above all the religious calendar. To this end, if we wish to pass beyond the specifically political-juridical interest of the corpus of texts constituting the *fasti*, that corpus must be augmented with a number of other types of source, most particularly the *ferialia*, the lists of feast days, and antiquarian literature.

In the light of this notion of the independent life of the published calendar, the *fasti*, the wall calendar of Marcus Fulvius Nobilior and Quintus Ennius acquires new significance. These particular *fasti* were influenced by Greek models, and were established in the Temple of the Muses, in the writers' assembly hall. They were not intended as a practical calendar; their purpose was historical. The fact that they were augmented with a list of consuls and censors made it possible to speak of '*fasti*' on a quite undifferentiated basis, whether one meant a calendar or a magistrate list. Against this background, the listing of the *dies natales* of temples, founded by victorious military commanders in particular, constituted a central building block in Fulvius' project of updating and historicizing the calendar; together with the restoring of responsibility for intercalation to the *pontifices*, effected a short time previously by the *lex Acilia*, this function also contributed to a sacralization of the *fasti* that was not without significance when it came to the addition of *feriae* in the period of Caesar and Augustus.

That it was this singular form of *fasti* that became the model for all later versions of the calendar says something about the historical conceptions of second-century Romans, as represented by Ennius; conceptions that, for example in the choice of coin imagery, and in the character of the Romulus-Quirinus figure, gave clear further impetus to the historicization of religious dates.[4] But the reception of these same Fulvian *fasti* also indicates that the everyday, practical significance of the existence of written calendars must not be overestimated: a large number of important *feriae indictivae* and *imperativae*, regular and exceptional holidays, evaded prescriptive written representation. Not until the Imperial Period do we find outside Rome isolated examples that depart from the canonic form, either by addition (Fasti Praenestini) or by subtraction (Fasti Guidizzolenses). The *ferialia*, tailored to the use of a particular group or locality, did not develop as a reduced form of *fasti*; the form they found was distinctive, and reminiscent of Greek examples of 'sacrifice calendars'.

In the Augustan period, both the relationship to the public realm and the independence of the *fasti* acquired a new dimension. The published *fasti*, which had begun to fill up with 'Imperial' festivals and dedications during Caesar's final months, and so had really become 'Augustan *fasti*' at that time, gained new political worth, irrespective of whether the demand for calendar-related information had been sated by book calendars, or was inexistent owing to geographical distance from Rome: they signified connection with the *consensus omnium* towards

[4] See e.g. KRÄMER 1965.

religious institutions in their traditional and renewed forms, and, in a particular way, brought loyalty to the *princeps* to the centre of that consensus. Along with its content, the medium itself moved into prominence, in the form of ever larger marble tablets.

This process did not begin at a central locality, such as a forum in the centre of Rome, but in the sanctuary of Dea Dia at the city's periphery. The audience addressed was primarily the college of the Arval brethren itself. From a Republican point of view, the cult can only be described as marginal; at least at the very inception of its transformation into a central element of the Imperial cult, a phase to which belong the *fasti* and the beginning of the lists in the fashion of *commentarii*, it needed the constant self-assurance that the marble furnishings and the integrative symbol of the calendar were admirably capable of supplying. The chronology of examples tell us that this was the starting point for the triumphal progress described by the marble *fasti* into the Tiberian period.

Such an emblematic use of the calendar, focussed not on the superficial level of its informative content, but on the symbolic message of the entire document, took it to places throughout the Italic world and its immediate vicinity. The requirement in the *lex Flavia municipalis*, that the Imperial festivals should be observed at a local level, an obligation that appears to have applied beyond the circle of towns subject to Latin law, explains this reception process only insofar as it demonstrates the Imperial cult to have been the bond that united an Empire-wide religion. It explains neither the complete adoption of the *fasti* formula developed in the city of Rome – the Feriale Cumanum illustrates the more specific alternative – nor the frequently private or collegiate initiatives for the erection of *fasti*, nor the genre's confinement to the Italic region. The last-mentioned factor may perhaps be explained by the still limited spread of the technical system underlying the calendar of the city of Rome (or closely analogous systems at least) at the crucial phase of the Early Principate.

One consequence of the increased public presence of *fasti* was the emergence of commentaries, which enhanced in turn the status of their subject, understood above all as 'ancient'. The typical shift of focus is seen in the attitudes of treatises on the subject: while Varro in his *Antiquitates rerum divinarum* (and also implicitly in *De lingua latina*) wrote *de feriis*, Ovid and, both before (literarily) and after him (epigraphically), Verrius Flaccus wrote works *de fastis*. Here, even in Roman literary discourse organizationally independent of the cult, we see the beginning of an important new genre.

Practical problems began to gain the upper hand in the period after Tiberius. The unrestricted increase in Imperial festivals reached its limits owing to financial consequences to the general economy (games cost money), and limitations to time in the courts. If the functional primacy of the Imperial festivals, whose impact was felt in the calendars of all factions in the Empire, was not to be undermined by divisive counter-strategies, of the kind that now restricted the generality of cults to specialists, it was necessary to limit the number of generally obligatory

Imperial festivals, and sharpen the profile of their content. Recognition of new dates was compensated by the deletion of old festivals. This lowered the incentive to list a historical snapshot in the calendar for ever and a day; at the same time, and this was probably more important, other genres arose in the context of the constantly growing presence of inscriptions, offering a substantially more favourable ratio of self-representation to the representation of external factors such as names of gods and emperors, and general information.

The model of the Roman *fasti* nevertheless persisted, culminating in the form of the Christian *sanctorale*, as demonstrated most succinctly by the Kalendarium marmoreum in Naples. Influence such as this must not be underestimated. In historical retrospect, this apparently isolated phenomenon evolved in a space that was full of alternatives. Among the factors that explain the 'victory' of the calendar of the city of Rome, down to idiosyncratic details such as the unusual length of February, and the religiously and historically pregnant month-names Martius and Augustus, no small part was played by its technical superiority and its presence in the Imperial administration and the military. A necessary condition for both these factors was ease of graphic representation, and thus publication and propagation of the calendar. In this respect, by present understanding the *fasti* had no equal.

Implementation of the ultimately slight Gregorian Reform took three hundred years, in the realm of the Julian Calendar alone, and this at a time that already knew printing with movable characters. This calendar that goes back, in its original form, to the Early Republican Period, may not be the world's only calendar, but it is the world's own, and, in view of the potential transitional cost of a new reform, the consensus up to now has been that it should so remain.

Abbreviations[1]

AC	*Antiquité classique.* Louvain-la-Neuve.
AJPh	*American Journal of Philology.* Baltimore.
ANRW	*Aufstieg und Niedergang der römischen Welt: Geschichte und Kultur Roms im Spiegel der neuen Forschung.* Ed. HILDEGARD TEMPORINI AND WOLFGANG HAASE. Berlin: de Gruyter, 1972–.
ARW	*Archiv für Religionswissenschaft.* Leipzig.
AU	*Der altsprachliche Unterricht.* Stuttgart.
BJ	*Bonner Jahrbücher.* Bonn.
CIL	*Corpus Inscriptionum Latinarum.* Berlin: Akademie-Verlag.
CP	*Classical Philology.* Chicago.
CQ	*Classical Quarterly.* Oxford.
CR	*Classical Review.* Oxford.
EPRO	Études préliminaires aux religions orientales dans l'empire romain. Leiden: Brill. Ernout, Meillet: see ERNOUT AND MEILLET 1959–60.
FIR	Bruns: *Fontes Iuris Romani Antiqui.* 7th edition. Ed. OTTO GRADENWITZ. Tübingen: Mohr, 1909. epicorum et lyricorum praeter Ennium et Lucilium. Post Aemilium Baehrens iterum edidit
HSCPh	*Harvard Studies in Classical Philology.* Cambridge, MA.
HThR	*Harvard Theological Review.* Cambridge, MA.

[1] Abbreviations for the works of classical authors conform to the index volume of *Thesaurus Linguae Latinae* ([2]1990).

The Roman Calendar from Numa to Constantine: Time, History and the Fasti, First Edition. Jörg Rüpke.

HZ *Historische Zeitschrift*. Munich.

ILLRP *Inscriptiones latinae liberae rei publicae*. Ed. Attilio Degrassi. Florence: Nuova Italia, 1957.

ILS *Inscriptiones latinae selectae*. 1-3.2. Ed. Hermann Dessau. Berlin: Weidmann, 1892–1916.

JbAC *Jahrbuch für Antike und Christentum*. Münster.

JdI *Jahrbuch des Deutschen Archäologischen Instituts*. Berlin.

JRS *Journal of Roman Studies*. London.

MDAI R *Mitteilungen des Deutschen Archäologischen Instituts in Rom*. Heidelberg.

MEFRA *Mélanges d'Archéologie et d'Histoire de l'École française de Rome (Antiquité)*. Paris.

MH *Museum Helveticum*. Basel.

NSc *Notizie degli scavi di antichità*. Rome.

PBSR *Papers of the British School at Rome*. London.

RAC *Reallexikon für Antike und Christentum*. Ed. Theodor Klauser and Ernst Dassmann et al. Stuttgart: Hiersemann, 1950–.

RE *Paulys Realencyclopädie der Classischen Altertumswissenschaft. Neue Bearbeitung*. Ed. Georg Wissowa (Kroll, Mittelhaus and Ziegler). Stuttgart: Metzler (Druckenmüller), 1893–1980.

REA *Revue des Études Anciennes*. Talence.

REL *Revue des Études Latines*. Paris.

RHDFE *Revue historique de droit français et étranger*. Paris.

RhM *Rheinisches Museum*. Frankfurt a. M.

RHR *Revue de l'Histoire des Religions* Paris. R. P. Wright, *Roman Inscriptions of Britain 1: Inscriptions on Stone*. Oxford: Clarendon.

RRC See Crawford 1974.

RVV Religionsgeschichtliche Versuche und Vorarbeiten.

SMSR *Studi e materiali della storia delle religioni*. Rome.

TAPhA *Transactions of the American Philological Association*. Philadelphia.

TLE *Testimonia linguae Etruscae*. Ed. Massimo Pallottino. Biblioteca di Studi superiori 24. 2a ed. Florence: Nuova Italia, 1968.

YCS *Yale Classical Studies*. New Haven.

ZPE *Zeitschrift für Papyrologie und Epigraphik*. Bonn.

ZSavSt *Zeitschrift der Savigny-Stiftung für Rechtsgeschichte. Romanistische Abt.* Weimar; Cologne.

References

ABERSON, MICHEL 1994. *Temples votifs et butin de guerre dans la Rome republicaine*. Bibliotheca Helvetica Romana 26. Rome: Institut suisse de Rome.

ACHELIS, HANS 1929. *Der Marmorkalender in Neapel*. [Prog.] Zur Feier des Reformationsfestes und des Übergangs des Rektorats … [Universität Leipzig]. Leipzig: Edelmann.

ALFÖLDI, ANDREAS 1966. Die Alexandrinischen Götter und die Vota Publica am Jahresbeginn. *JbAC* 8–9, 53–87.

ALFÖLDY, GÉZA 1991. *Augustus und die Inschriften: Tradition und Innovation: Die Geburt der imperialen Epigraphik. Gymnasium* 98, 289–324.

ALTHEIM, FRANZ 1932. Minerva. 1. *RE* 5 (2), 1774–1802.

AMPOLO, CARMINE 1988. Lavinium and Rome: [review] M. Torelli … *CR* 38, 117–20.

ANDREAU, JEAN (ed.) 2000. Les marchés hebdomadaires du Latium et de Campanie au Ier siècle ap. J.-C. In ELIO LO CASCIO (ed.), *Mercati permanenti e mercati periodici nel mono romano.* Atti degli incontri capresi di storia dell'economia antica (Capri, October, 1997). Bari: Edipuglia, 69–91.

ARNOLD, DIETER 1992. *Die Tempel Ägyptens: Götterwohnungen, Kultstätten, Baudenkmäler.* Zürich: Artemis & Winkler.

AULARD, A[LPHONSE] 1927. *Christianity and the French Revolution*. Tr. from the French by Lady Frazer. London: Ernest Benn. [*Le christianisme et la révolution française.* Christianisme 7. Paris: Rieder, 1925.]

AUSBÜTTEL, FRANK M. 1982. *Untersuchungen zu den Vereinen im Westen des Römischen Reiches*. Frankfurter Althistorische Studien 11. Kallmünz: Laßleben.

BACCHIOCCHI, SAMUELE 1977. *From Sabbath to Sunday: A Historical Investigation of the Rise of Sunday Observance in Early Christianity*. Rome: Pontifical Gregorian University Press.

BACZKO, BRONISLAW 1984. Le calendrier républicain: Décréter l'éternité. In PIERRE NORA (ed.), *Les lieux de mémoire 1: La république*. Avec la collaboration de CHARLES-ROBERT AGERON. Paris: Gallimard, 37–83.

The Roman Calendar from Numa to Constantine: Time, History and the Fasti, First Edition. Jörg Rüpke.

BALDOVIN, JOHN FRANCIS 1987. *The Urban Character of Christian Worship: The Origins, Development, and Meaning of Stational Liturgy*. Orientalia Christiana analecta 228. Rome: Pontificium Institutum Studiorum Orientalium.

BALDOVIN, JOHN FRANCIS 1991. *Worship: City, Church and Renewal*. Washington, DC: Pastoral Press.

BASNIZKI, LUDWIG 1938. *Der jüdische Kalender: Entstehung und Aufbau*. [Reprint with postface; Frankfurt a. M.: Kauffmann, 1938.] Frankfurt a. M.: Athenäum.

BAUCKHAM, RICHARD J. 1982. The Lord's Day. In D. A. CARSON (ed.), *From Sabbath to Lord's Day: A Biblical, Historical, and Theological Investigation*. Grand Rapids: Zondervan, 221–50.

BAXMANN, INGE 1989. *Die Feste der Französischen Revolution: Inszenierung von Gesellschaft als Natur*. Ergebnisse der Frauenforschung 17. Weinheim: Beltz.

BEARD, MARY 1985. Writing and Ritual: A Study of Diversity and Expansion in the Arval Acta. *PBSR* 53, 114–62.

BEARD, MARY 1987. A Complex of Times: No More Sheep on Romulus' Birthday. *Proceedings of the Cambridge Philological Society* 33, 1–15.

BEARD, MARY 1991. Writing and Religion: Ancient Literacy and the Function of the Written Word in Roman religion. *Journal of Roman Archaeology* suppl. 3.

BEARD MARY et al. 1991. *Literacy in the Roman World*. *Journal of Roman Archaeology*, suppl. 3.

BEAUJEU, JEAN 1976. Les dernières années du calendrier pré-julien. *Mélanges offerts à Jacques Heurgon: L'Italie préromaine et la Rome républicaine*. Collection École française de Rome 27. Rome: École française.

BELLEN, HEINZ 1985. *Metus Gallicus – Metus Punicus: Zum Furchtmotiv in der römischen Republik*. Akademie der Wissenschaften und Literatur, Abhandlungen der geistes- und sozialwissenschaftlichen Klasse 1985,3. Mainz / Stuttgart: Akademie / Steiner.

BELOCH, KARL JULIUS 1918. Der römische Kalender von 218–168. *Klio* 15, 382–419.

BELOCH, KARL JULIUS 1929. Der römische Kalender Varr. 565 und 566 (189 v. Chr.). *Klio* 22, 464–6.

BEN-DOV, JONATHAN 2008. *Head of All Years: Studies in Qumran Calendars and Astronomy in their Ancient Background*. Studies in the Texts from the Desert of Judah 78. Leiden: Brill.

BENVENISTE, ÉMILE 1969. *Le vocabulaire des institutions indo-européennes 2: Pouvoir, droit, religion*. Sommaires, tableau et index établis par JEAN LALLOT. Paris: Éditions de Minuit. [German: *Indoeuropäische Institutionen: Wortschatz, Geschichte, Funktionen*, Frankfurt a. gM. / Paris: Campus / Éditions de la Maison des Sciences de l'Homme, 1993.]

BERGMANN, WERNER 1984. Der römische Kalender: Zur sozialen Konstruktion der Zeitrechnung. Ein Beitrag zur Soziologie der Zeit. *Saeculum* 35, 1–15.

BISCHOFF, ERNST 1884. De Fastis Graecorum antiquioribus. *Leipziger Studien* 7, 313–416.

BISCHOFF, ERNST 1919. Kalender (griechischer). *RE* 10 (2), 1568–1602.

BLEICKEN, JOCHEN 1955. *Das Volkstribunat der klassischen Republik: Studien zu seiner Entwicklung zwischen 287 und 133 v. Chr.* 2. [11955.] Zetemata 13. Munich: Beck.

BLEICKEN, JOCHEN 1981. Zum Begriff der römischen Amtsgewalt: auspicium – potestas – imperium. *Nachrichten der Akademie der Wissenschaften in Göttingen, phil.-hist. Kl. 1981,9*. Göttingen: Vandenhoeck & Ruprecht, 255–300.

BÖMER, FRANZ 1958. *P. Ovidius Naso, Die Fasten 2: Kommentar.* Heidelberg: Winter.

BONNEFOND-[COUDRY], MARIANNE 1989. *Le sénat de la république romaine de la guerre d'Hannibal à Auguste: Pratiques délibératives et prise de décision.* Bibliothèque des Écoles françaises d'Athènes et de Rome 273. Rome: École française.

BORGEAUD, WILLY ALFRED 1982. *Fasti Umbrici: Études sur le vocabulaire et le rituel des Tables eugubines.* Collection d'études anciennes de l'Université d'Ottawa 1. Ottawa: Édition de l'Université.

BOSWORTH, A. B. 1982. Augustus and August: Some Pitfalls of Historical Fiction. *HSCPh* 86, 151–70.

BOURDIEU, PIERRE 1990. *The Logic of Practice.* Tr. from the French by Richard Nice. Cambridge: Polity Press. [*Le sens pratique.* Paris: Les Éditions de Minuit; Maison des Sciences de l'Homme, 1980.]

BOWMAN, ALAN K. 1991. Literacy in the Roman Empire: Mass and Mode. In BEARD et al. 1991, 119–31.

BOYANCÉ, PIERRE 1955. *Fulvius Nobilior et le dieu ineffable. RPh* 29, 172–92.

BOYLE, A. J. AND DOMINIK, J. W. (eds) 2003. *Flavian Rome: Culture, Image, Text.* Leiden: Brill.

BREMMER, JAN N. 1987. Romulus, Remus and the Foundation of Rome. In NICHOLAS M. HORSFALL (eds), *Roman Myth and Mythography.* BICS, Suppl. 52. London: Institute of Classical Studies, 25–48.

BREMMER, JAN N. 1990. Adolescents, Symposion, and Pederasty. In OSWYN MURRAY (ed.). *Sympotica: A Symposium on the Symposion.* Oxford: Clarendon, 135–48.

BRIND'AMOUR, PIERRE 1983. *Le calendrier romain: Recherches chronologiques.* Collection d'études anciennes de l'Université d'Ottawa 2. Ottawa: Éditions de l'Université d'Ottawa.

BRISCOE, JOHN 1991. [Review] Gerhard Radke: Fasti Romani ... 1990 ... *CR* 41, 404–6.

BROUGHTON, T. R. S. 1951–2. *The Magistrates of the Roman Republic.* 2 vols. American Philological Association, Philological Monographs 15, 1–2. New York: American Philological Association.

BROUGHTON, T. R. S. 1986 [1960]. *The Magistrates of the Roman Republic III: Supplement.* 2nd edn. American Philological Association Monographs 15. Atlanta: Scholars Press.

BROWN, PETER 1981. *The Cult of the Saints: Its Rise and Function in Latin Christianity.* [*Die Heiligenverehrung,* Leipzig, 1991.] Haskell Lectures on History of Religions N. S. 2. Chicago: Chicago University Press.

BURCK, ERICH 1992. *Das Geschichtswerk des Titus Livius.* Bibliothek der klassischen Altertumswissenschaft R. 2, 87. Heidelberg.

BURCKHARDT, LEONHARDT ALEXANDER 1988. *Politische Strategien der Optimaten in der späten römischen Republik.* Historia Einzelschriften 57. Stuttgart: Steiner.

BUREN, ALBERT WILLIAM van 1923. Graffiti ad Ostia. *CR* 37, 163–4.

BURKERT, WALTER 1961. Hellenistische Pseudopythagorica. *Philologus* 105, 16–43, 226–46.

BURKERT, WALTER 1985 *Greek Religion, Archaic and Classical.* Tr. from the German by JOHN RAFFAN. Oxford: Blackwell. [*Griechische Religion der archaischen und klassischen Epoche.* Die Religionen der Menschheit 15. Stuttgart: Kohlhammer, 1977.]

CAMASSA, GIORGIO 1984. Sull'origine e le funzioni del culto di Volcanus a Roma. *Rivista storica Italiana* 96, 811–54.

CAMERON, ALAN 1990. Forschungen zum Thema der 'heidnischen Reaktion' in der Literatur seit 1943. In ANDREAS ALFÖLDI AND ELISABETH ALFÖLDI (eds), *Die*

Kontorniat-Medallions 2: Text. Antike Münzen und Geschnittene Steine 6, 2. Berlin: de Gruyter, 63–74.

CANCIK, HUBERT 1986. Nutzen, Schmuck und Aberglaube: Ende und Wandlungen der römischen Religion im 4. und 5. Jahrhundert. In HARTMUT ZINSER (ed.), *Der Untergang von Religionen*. Berlin: Reimer, 65–90.

CAVAIGNAC, EUGÈNE 1915. La chronologie romaine de 215 à 168. *Klio* 14, 37–42.

CHANIOTIS, ANGELOS 1991. Gedenktage der Griechen: Ihre Bedeutung für das Geschichtsbewußtsein griechischer Poleis. In Assmann (ed.), 1991, 123–45.

CHASTAGNOL, ANDRÉ 1987. Aspects concrets et cadre topographique des fêtes décennales des empereurs à Rome. *L'Urbs: Espace urbain et histoire (Ier siècle av. J.-C.–IIe siècle ap. J.-C.)*. Actes du colloque international ... 8–12 May 1985. Rome: École française.

CHASTAGNOL, ANDRE 1988a. Les fêtes décennales des empereurs à Rome. In P. GIGNOUX (ed.), *La commémoration: Colloque du centenaire de la Section des Sciences Religieuses de l'École Pratique des Hautes Études*. Louvain, 107–14.

CHASTAGNOL, ANDRE 1988b. Les inscriptions des monuments inaugurés lors des fêtes impériales. *MEFRA* 100, 13–26.

CHUVIN, PIERRE 1990. *A Chronicle of the Last Pagans*. Tr. by B. A. ADO. Revealing Antiquity Series 4. Cambridge, MA: Harvard University Press.

CLAVEL-LÉVÊQUE, MONIQUE 1984. *L'Empire en jeux: Espace symbolique et pratique sociale dans le monde romain*. Paris: CNRS.

CLAVEL-LÉVÊQUE, MONIQUE 1986. L'espace des jeux dans le monde romain: hégémonie, symbolique et pratique sociale. *ANRW II* 16 (3), 2405–563.

COARELLI, FILIPPO 1972. Il sepolcro degli Scipioni. *Dialoghi d'Archeologia* 6, 36–106.

COARELLI, FILIPPO 1975. *Rom: Ein archäologischer Führer*. [*Guida Archeologica di Roma*, 1974.] Freiburg i. Br.: Herder.

COLSON, FRANCIS HENRY 1926. *The Week: An Essay on the Origin and Development of the Seven-day Cycle*. Cambridge: University Press.

CONNOR, W. R. 1987. Tribes, Festivals and Processions: Civic Ceremonial and Political Manipulation in Archaic Greece. *JHS* 107, 40–50.

CORBIER, MIREILLE 1991. L'écriture en quête de lecteurs. In Beard et al. (eds), 1991. 99–118.

CRAWFORD, MICHAEL HEWSON 1974. *Roman Republican Coinage*. 2 vols. London: Cambridge University Press.

CRISTOFANI, M. 1995. *Tabula Capuana: Un calendario festivo di età arcaica*. Florence.

DEGRASSI, ATTILIO (ed.) 1947. *Inscriptiones Italiae 13: Fasti et elogia. Fasciculus 1: Fasti consulares et triumphales*. 2 parts. Academiae Italicae consociatae ediderunt. Rome: Libreria dello stato.

DEGRASSI, ATTILIO (ed.) 1963. *Inscriptiones Italiae 13: Fasti et elogia. Fasciculus 2: Fasti anni Numani et Iuliani, accedunt ferialia, menologia rustica, parapegmata*. 2 parts. Academiae Italicae consociatae ediderunt. Rome: Libreria dello stato.

DELEHAYE, HIPPOLYTE 1933. *Les origines du culte des martyrs*. 2nd edn, rev. Subsidia Hagiographica 20. Brussels: Bollandistes.

DEMANDT, ALEXANDER 1989. *Die Spätantike: Römische Geschichte von Diocletian bis Justinian 284–565 n. Chr*. Handbuch der Altertumswissenschaft 3, 6. Munich: Beck.

DE MARTINO, FRANCESCO 1985. *Wirtschaftsgeschichte des alten Rom*. German translation from the Italian by BRIGITTE GALSTERER. Munich: Beck [1979-80: *Storia economica di Roma*. Firenze: *La nuova Italia*].

DENIS-BOULET, FORMERLY MAURICE-DENIS, NOËLE 1960. *The Christian Calendar*. Tr. from the French by P. J. HEPBURNE-SCOTT. London: Burns & Oates. [*Le calendrier chrétien*, Paris: Fayard, 1959.]

DERCHAIN, PHILIPPE 1962. Mythes et dieux lunaires en Égypte. *La lune, mythes et rites: Égypte, Sumer, Babylone, Cambodge, Japon, Chine, Sibérie*. Sources Orientales 5. Paris: Éditions du Seuil.

DEROW, P. S. 1976. The Roman Calendar 218–191 BC. *Phoenix* 30, 265–81.

DESSAU, HERMANN 1900. Zum Kalender der Provinz Asien. *Hermes* 35, 332–8.

DEUBNER, LUDWIG 1911. Zur Entwicklungsgeschichte der altrömischen Religion. *Neue Jahrbücher für das klassische Altertum* 27 (1), 321–35. [Reprint Deubner 1982, 113–27.]

DIHLE, ALBRECHT 1980. Zur spätantiken Kultfrömmigkeit. In JOCHEN MARTIN AND BARBARA QUINT (eds), *Christentum und antike Gesellschaft*. [JbAC-Ergbd. 18, 1980, 39–54.] Wege der Forschung 649. Darmstadt: Wissenschaftliche Buchgesellschaft, 143–68.

DINZELBACHER, PETER 1990. Die 'Realpräsenz' der Heiligen in ihren Reliquiaren und Gräbern nach mittelalterlichen Quellen. In DIETER R. BAUER (ed.), *Heiligenverehrung in Geschichte und Gegenwart*. Ostfildern: Schwabenverlag, 115–74.

D'IPPOLITO, FEDERICO 1985. Das ius Flavianum und die lex Ogulnia. *ZSavSt* 102, 91–128.

D'IPPOLITO, FEDERICO 1986. *Giuristi e sapienti in Roma arcaica*. Rome: Laterza.

Divination 1966. *La divination en Mésopotamie ancienne et dans les régions voisines: XIVe Rencontre Assyriologique Internationale*. Paris: Presses Universitaires.

Divination 1982. *Divination: Religionshistorisk Forenings Symposium 20–1. Januar 1982 i København*. Udgivet af Religionshistorisk Forening. Scernummer af Chaos. Copenhagen.

DIX, GREGORY 1945. *The Shape of the Liturgy.* 4th impr. of 2nd edn. London: Dacre.

DÖLGER, FRANZ JOSEPH 1940. Die Planetenwoche der griechisch-römischen Antike und der christliche Sonntag. *Antike und Christentum* 6, 202–38; Plates 4–7.

DOUGLAS, [Tew] MARY 1973 [1970]. *Natural Symbols: Explorations in Cosmology*. London: Barrie & Jenkins. 2nd edn, 1973.

DOW, STERLING 1961. The Walls Inscribed with Nikomakhos' Law Code. *Hesperia* 30, 58–73, pl. 9–11.

DRUMANN, WILHELM 1906. *Geschichte Roms in seinem Übergange von der republikanischen zur monarchischen Verfassung oder Pompeius, Caesar, Cicero und ihre Zeitgenossen nach Geschlechtern und mit genealogischen Tabellen 3: Domitii-Julii*. 2nd edition, ed. P. GROEBE. Leipzig: Bornträge.

DULABAHN, ELIZABETH SNOWDEN 1987. Studies on the 'Laterculus' of Polemius Silvius. Diss. Bryn Mawr College.

DURKHEIM, ÉMILE 1915. *The Elementary Forms of the Religious Life: A Study in Religious Sociology*. Tr. from the French by JOSEPH WARD SWAIN, 1915. Repr. Glencoe, IL: Free Press, 1947.

DUVAL, YVES-MARIE 1977. Des Lupercales de Constantinople aux Lupercales de Rome. *REL* 55, 222–70.

DUVAL, YVETTE 1982. *Loca sanctorum Africae: Le culte des martyrs en Afrique du IVe au VIIe siècle*. 2 vols. Collection de l'École française de Rome 58. Rome: École française.

ECK, WERNER 1984. Senatorial Self-representation: Developments in the Augustan Period. In FERGUS MILLAR AND ERICH SEGAL (eds), *Caesar Augustus: Seven Aspects*. Oxford: Clarendon.

EDLUND-BERRY, INGRID E. M. 1992. Etruscans at Work and Play: Evidence for an Etruscan Calendar. In H. FRONING, TONIO HÖLSCHER AND HARALD MIELSCH (eds), *Kotinos: Festschrift für Erika Simon*. Mainz: Zabern, 330–8.

EHRENBERG, VICTOR; JONES, A. H. M. 1976. *Documents illustrating the reigns of Augustus and Tiberius*. Repr. of 2nd edn with addenda [^{2}1955]. Oxford: Oxford University Press.

EHRHARDT, NORBERT 1984. Ein milesischer Festkalender aus severischer Zeit. *MDAI Istanbul* 34, 371–404.

ELIAS, NORBERT 2007. *An Essay on Time*. In Elias, *The Collected Works of Norbert Elias*, 9. Dublin: University College Dublin Press. Tr. from the German. [1988. *Über die Zeit: Arbeiten zur Wissenssoziologie* 2. Ed. von MICHAEL SCHRÖTER. Frankfurt a. M.: Suhrkamp.]

ELM VON DER OSTEN, DOROTHEE, RÜPKE, JÖRG AND WALDNER, KATHARINA (eds) 2006. *Literatur als Medium und Reflexion von Religion im römischen Reich* (Potsdamer altertumswissenschaftliche Beiträge 14). Stuttgart: Steiner.

ERNOUT, ALFRED AND MEILLET, ANTOINE 1959–60. *Dictionnaire étymologique de la langue latine: Histoire des mots*. 2 vols. 4th edn, rev., corr., augment. with index. Paris: Klincksieck.

EVANS, J. A. S. 1961. A Social and Economic History of an Egyptian Temple in the Greco-Roman Period. *YCS* 17, 143–283.

EVANS-PRITCHARD, E. E. 1939. Nuer Time-reckoning. *Africa* 12, 189–216.

FANTHAM, ELIZABETH 2002. Ovid's Fasti: Politics, History, and Religion. In BARBARA WEIDEN BOYD (ed.), *Brill's Companion to Ovid*. Leiden: Brill, 197–233.

FEENEY, DENIS 1998. *Literature and Religion at Rome: Cultures, Contexts, and Beliefs*. Cambridge: Cambridge University Press.

FEENEY, DENIS 2007. *Caesar's Calendar: Ancient Time and the Beginnings of History*. Berkeley: University of California Press.

FELDMAN, LOUIS H. 1993. *Jew and Gentile in the Ancient World: Attitudes and Interactions from Alexander to Justinian*. Princeton: Princeton University Press.

FERRARY, JEAN-LOUIS 1988. *Philhellénisme et impérialisme: Aspects idéologiques de la conquête romaine du monde hellénistique, de la seconde guerre de Macédoine à la guerre contre Mithridate*. Bibliothèque des Écoles françaises d'Athènes et de Rome 271. Rome: École française.

FÉVRIER, PAUL-ALBERT 1977. À propos du repas funéraire: Culte et sociabilité. *Cahiers archéologiques* 26, 29–45.

FLAIG, E. (2003). *Ritualisierte Politik: Zeichen, Gesten und Herrschaft im Alten Rom. Historische Semantik 1*. Göttingen: Vandenhoeck & Ruprecht.

FLAMANT, JACQUES 1984. L'année lunaire aux origines du calendrier pré-julien. *MEFRA* 96, 175–93.

FLEX, RUDOLF 1880. 'Die älteste Monatseintheilung der Römer.' Diss. Jena.

FONTAINE, JACQUES 1982. 'Le culte des saints et ses implications sociologiques: Réflexions sur un récent essai de Peter Brown.' *Acta Bollandiana* 100, 17–41.

FÖRSTER, HANS 2000. *Die Feier der Geburt Christi in der Alten Kirche: Beiträge zur Erforschung der Anfänge des Epiphanias- und Weihnachtsfests*. Studien und Texte zu Antike und Christentum 4. Tübingen: Mohr Siebeck.

Förster, Hans 2007. *Die Anfänge von Weihnachten und Epiphanias: Eine Anfrage an die Entstehungshypothesen*. Studien und Texte zu Antike und Christentum 46. Tübingen: Mohr Siebeck.

Fowler, William Warde 1969 [1899]. *The Roman Festivals of the Period of the Republic: An Introduction to the Study of the Religion of the Romans*. Repr. Port Washington: Kennikat.

Fraschetti, Augusto 1985. Cesare e Antonio ai Lupercalia. In Frederick Mario Fales and Cristiano Grottanelli (eds), *Sopranaturale e potere: nel mondo antico e nelle società tradizionali*. Milan: Angeli.

Fraschetti, Augusto 1988. 'Commemorare il principe: Ricerche preliminari sui calendari di epoca augusteo-Tiberiana.' In Philippe Gignoux (ed.), *La commémoration: Colloque du centenaire de la Section des Sciences Religieuses de l'École Pratique des Hautes Études*. Bibliothèque de l'École Pratique des Hautes Études, Sciences Religieuses 91. Louvain: Peeters, 115–40.

Fraschetti, Augusto 1990. 'Cognata numina': Culti della città e culti della famiglia del principe in epoca augustea. In Jean Andreau and Hinnerk Bruhns (eds), *Parenté et stratégies familiales dans l'antiquité romaines*. Actes de la table ronde des 2–4 octobre 1986. Collection de l'École française de Rome 129. Rome: École française, 85–119.

Fraser, Julius T. 1987. *Time: The Familiar Stranger*. Amherst: University of Massachusetts Press.

Frazer, James George 1929. *Publii Ovidii Nasonis Fastorum Libri Sex: The Fasti of Ovid*. 5 vols. London: Macmillan.

Frei-Stolba, Regula 1967. *Untersuchungen zu den Wahlen in der römischen Kaiserzeit*. Zürich: Juris.

Frere, Walter Howard 1930. *Studies in Early Roman Liturgy 1: The Kalendar*. Alcuin Club Collections 28. Oxford: Oxford University Press,.

Friedländer, Ludwig 1920 [1908]. *Darstellungen aus der Sittengeschichte Roms in der Zeit von Augustus bis zum Ausgang der Antonine 2*. 9th revised and augmented edn by Georg Wissowa. Leipzig: Hirzel. [Earlier edn, *Roman Life and Manners under the Early Empire*. Tr. by L. A. Magnus. Freese. London: Routledge, 1908.]

Gagé, Jean 1935. *Res gestae divi Augusti ex monumentis Ancyrano et Antiocheno Latinis Ancyrano et Apolloniensi Graecis*. Texte établi et comm. par ... 2ème éd. Repr. 1935. Publications de la faculté de lettres de l'Université de Strasbourg: Textes d'études 5. Paris: Belles Lettres, 1950.

Galil, Gershon 1991. 'The Babylonian Calendar and the Chronology of the Last Kings of Judah.' *Biblica* 72, 367–78.

Gamber, Klaus 1956. *Wege zum Urgregorianum: Erörterung der Grundfragen und Rekonstruktionsversuch des Sakramentars Gregors d. Gr. vom Jahre 592*. In Verbindung mit Alban Dold ed. von ... Texte und Arbeiten R. 1, 46. Beuron: Kunstverlag.

Gamber, Klaus 1970. *Missa Romensis: Beiträge zur frühen römischen Liturgie und zu den Anfängen des Missale Romanum*. Studia Patristica et Liturgica 3. Regensburg: Pustet.

Gandz, Solomon 1949. 'The Origin of the Planetary Week or the Planetary Week in Hebrew Literature.' *Proceedings of the American Academy for Jewish Research* 18, 213–54.

Gee, Emma 2000. *Ovid, Aratus and Augustus: Astronomy in Ovid's Fasti*. Cambridge: University Press.

Gesche, Helga 1976. *Caesar*. Erträge der Forschung 51. Darmstadt: Wissenschaftliche Buchgesellschaft.

GIESECKE, MICHAEL 1991. *Der Buchdruck in der frühen Neuzeit: Eine historische Fallstudie über die Durchsetzung neuer Informations- und Kommunikationstechnologien.* Frankfurt a. M.: Suhrkamp.

GJERSTAD, EINAR 1961. 'Notes on the Early Roman Calendar.' *Acta Archaeologica (Kopenhagen)* 32, 193–214.

GLADIGOW, BURKHARD 1992. 'Audi Juppiter, Audite Fines.' In OKKO BEHRENDS AND LUIGI CAPOGROSSI COLOGNESI (eds), *Die römische Feldmeßkunst: Interdisziplinäre Beiträge zu ihrer Bedeutung für die Zivilisationsgeschichte Roms.* Göttingen: Vandenhoeck & Ruprecht, 172–91.

GOLDENBERG, ROBERT 1979. 'The Jewish Sabbath in the Roman World up to the Time of Constantine the Great.' *ANRW II* 19 (1), 414–47.

GOLDSTEIN, BERNARD R. AND COOPER, ALAN 1990. 'The Festivals of Israel and Judah and the Literary History of the Pentateuch.' *Journal of the American Oriental Society* 110, 19–31.

GONZÁLEZ, JULIAN 1986. 'The Lex Irnitana: A New Copy of the Flavian Municipal Law.' *JRS* 76, 147–243.

GOODY, JACK 1968. 'Time II: Social Organization.' In DAVID C. SILLS (ed.), *International Encyclopedia of the Social Sciences.* 16. New York: Macmillan, 30–42.

GOODY, JACK 1986. *The Logic of Writing and the Organization of Society.* Cambridge: Cambridge University Press.

GORDON, ARTHUR E. 1983. *Illustrated Introduction to Latin Epigraphy.* Berkeley: University of California Press.

GRAF, FRITZ 1985. *Nordionische Kulte: Religionsgeschichtliche und epigraphische Untersuchungen zu den Kulten von Chios, Erythrai, Klazomenai und Phokaia.* Bibliotheca Helvetica Romana 21. Rome: Istituto Svizzero di Roma.

GRAF, FRITZ 1993a. Einleitung. In GRAF (ed.), *Mythos in mythenloser Gesellschaft: Das Paradigma Roms.* Colloquium Rauricum 3. Stuttgart: Teubner, 1–5.

GRAF, FRITZ 1993b. Der Mythos bei den Römern: Forschungs- und Problemgeschichte. In Graf (ed.) (1993), 25–43.

GRAFTON, ANTHONY T. AND SWERDLOW, NOEL M. 1988. 'Calendar Dates and Ominous Days in Ancient Historiography.' *Journal of the Warburg and Courtauld Institutes* 51, 14–42.

GREEN, STEVEN J. 2004. *Ovid, Fasti 1: A Commentary.* Mnemosyne suppl. 251. Leiden: Brill.

GROS, PIERRE 1976. *Aurea Templa: Recherches sur l'architecture religieuse de Rome à l'époque d'Auguste.* Bibliothèque des Écoles françaises d'Athènes et de Rome 231. Rome: École française.

GROS, PIERRE AND SAURON, GILLES 1988. Das politische Programm der öffentlichen Bauten. In Antiken-Museum Berlin (ed.), *Kaiser Augustus und die verlorene Republik.* Berlin: Staatliche Museen Preußischer Kulturbesitz, 48–68.

GRUEN, ERICH S. 1984. *The Hellenistic World and the Coming of Rome.* 2 vols. Berkeley: University of California Press.

GRUEN, ERICH S. 1992. *Culture and National Identity in Republican Rome.* Cornell Studies in Classical Philology 52. Ithaca, NY: Cornell University Press.

GRZYBEK, ERHARD 1990. *Du calendrier macédonien au calendrier ptolémaïque: Problèmes de chronologie hellénistique.* Schweizerische Beiträge zur Altertumswissenschaft 20. Basel: Reinhardt.

GUTBERLET, DAGMAR 1985. *Die erste Dekade des Livius als Quelle zur gracchischen und sullanischen Zeit*. Beiträge zur Altertumswissenschaft 4. Hildesheim: Olms-Weidmann.

HACKL, URSULA 1982. *Senat und Magistratur in Rom von der Mitte des 2. Jahrhunderts v. Chr. bis zur Diktatur Sullas*. Regensburger historische Forschungen 9. Kallmünz: Laßleben.

HALKIN, LÉON 1953. *La supplication d'action de graces chez les Romains*. Bibliothèque de la Faculté de Philosophie et Lettres de l'Université de Liège 128. Paris: Belles Lettres.

HANNAH, ROBERT 2005. *Greek and Roman Calendars: Constructions of Time in the Classical World*. London: Duckworth.

HANNESTAD, NIELS 1986. *Roman Art and Imperial Policy*. Jutland Archaeological Society Publications 19. Århus: Aarhus University Press.

HARE, JULIUS C. 1832. 'On the Names of the Days of the Week.' *Philological Museum* 1–73.

HARNONCOURT, PHILIPP 1985. Osterkomputation – Geschichtstheologie – Theologiegeschichte. *Archiv für Liturgiewissenschaft* 27, 263–72.

HARRIS, WILLIAM VERNON 1989. *Ancient Literacy*. Cambridge, MA: Harvard University Press.

HARTMANN, OTTO ERNST 1882. *Der römische Kalender*. Aus dem Nachlasse des Verfassers ed. von LUDWIG LANGE. Leipzig: Teubner.

HASELBERGER, LOTHAR 2008 [2002]. *Mapping Augustan Rome*. Portsmouth: *Journal of Roman Archaeology*, suppl. 50, rev. edn 2008.

HEIM, FRANÇOIS 1992. *La théologie de la victoire de Constantin à Théodose*. Théologie historique 89. Paris: Beauchesne.

HEINEMANN, KLAUS AND LUDES, PETER 1978. Zeitbewußtsein und Kontrolle der Zeit. In KURT HAMMERICH AND MICHAEL KLEIN (eds), *Materialien zur Soziologie des Alltags*. Kölner Zeitschrift für Soziologie und Sozialpsychologie: Sonderheft 20. Opladen: Westdeutscher Verlag, 220–43.

HEINEN, HUBERT 1910. *Zur Begründung des römischen Kaiserkultes: Chronologische Übersicht von 48 v. bis 14 n. Chr.* Tübingen: Laupp.

HENNIG, JOHN 1980. Kalendar und Martyrologium als Literaturformen. In HENNIG (ed.), *Literatur und Existenz: Ausgewählte Aufsätze*. Heidelberg: Winter, 37–80.

HENSCHEL, OSKAR 1963. Quinquatrus. *RE* 24, 1149–62.

HERMANS, M. E. H. 1964. Endotercisus. *MH* 21, 173–6.

HERMANSEN, GUSTAV 1940. *Studien über den italischen und den römischen Mars*. Tr. GLÖDE. Copenhagen: Gyldendal.

HERZ, PETER 1975. Untersuchungen zum Festkalender der römischen Kaiserzeit nach datierten Weih- und Ehreninschriften. 2 vols. Diss. Mainz, 1–403, 404–604.

HERZ, PETER 1978. Kaiserfeste der Prinzipatszeit. *ANRW II* 16 (2), 1135–1200.

HEURGON, JACQUES 1957. *Trois études sur le 'Ver sacrum'*. Collection Latomus 26. Brussels: Latomus.

HEUSS, ALFRED 1944. Zur Entwicklung des Imperiums der römischen Oberbeamten. *ZSavSt* 64, 57–133.

HEUSS, ALFRED 1976. *Römische Geschichte*. 4th augmented edition. Braunschweig: Westermann.

HEUSS, ALFRED 1982. *Gedanken und Vermutungen zur frühen römischen Regierungsgewalt*. Nachrichten aus der Akademie der Wissenschaften in Göttingen 1982, phil.-hist. Kl. Göttingen: Vandenhoeck & Ruprecht, 375–454.

HÖLKESKAMP, KARL-JOACHIM 1987. *Die Entstehung der Nobilität: Studien zur sozialen und politischen Geschichte der Römischen Republik im 4. Jhdt. v. Chr.* Stuttgart: Steiner.

HÖLKESKAMP, KARL-JOACHIM 1988. Das plebiscitum Ogulnium de sacerdotibus: Überlegungen zu Authentizität und Interpretation der livianischen Überlieferung. *RhM* 131, 51–67.

HÖLKESKAMP, KARL-JOACHIM 1988a. Die Entstehung der Nobilität und der Funktionswandel des Volkstribunats: Die historische Bedeutung der lex Hortensia de plebiscitis. *Archiv für Kulturgeschichte* 70, 271–312.

HÖLSCHER, TONIO 1980. Die Geschichtsauffassung in der römischen Repräsentationskunst. *JdI* 95, 265–321.

HÖLSCHER, TONIO 1984. *Staatsdenkmal und Publikum: Vom Untergang der Republik bis zur Festigung des Kaisertums in Rom.* Xenia 9. Konstanz: Universitätsverlag.

HOHL, ERNST 1942. Das Angebot des Diadems an Cäsar. *Klio* 34, 92–117.

HOLLAND, LOUISE ADAMS 1953. Septimontium or Saeptimontium. *TAPhA* 84, 16–34.

HOLLEMAN, A. W. J. 1976. *Pope Gelasius I and the Lupercalia.* Amsterdam: Hakkert.

HOLLEMAN, A. W. J. 1978. Cicero's Reaction to the Julian Calendar. *Historia* 27, 496–8.

HUMM, MICHEL (2005) *Appius Claudius Caecus: La République accomplie.* Rome: Ecole Française de Rome.

IDELER, LUDWIG 1825, 1826. *Handbuch der mathematischen und technischen Chronologie.* 2 vols. Berlin: Rücker.

ILYAS, MOHAMMAD 1982. New Moon's First Visibility: Review of Astronomy and Current Islamic Calendrical Practices. *Islamic Culture* 56, 43–63.

JACQUES, FRANÇOIS AND SCHEID, JOHN 1990. *Rome et l'intégration de l'empire (44 av. J. gC.-260 ap. J. C.) 1: Les structures de l'Empire romain.* 2nd edn. Paris: Presses Universitaires.

JEHNE, MARTIN 1987. *Der Staat des Dictators Caesar.* Passauer historische Forschungen 3. Cologne: Böhlau.

JOCELYN, H. D. 1966. The Roman Nobility and the Religion of the Republican State. *Journal of Religious History* 4, 89–104.

JOHNEN, CHRISTIAN 1911. *Geschichte der Stenographie im Zusammenhang mit der allgemeinen Entwicklung der Schrift und der Schriftkürzung 1: Die Schriftkürzung und Kurzschrift im Altertum, Mittelalter und Reformationszeitalter ...* Berlin: Schrey.

JONES, CHARLES W. (ed.) 1943. *Bedae, Opera de temporibus.* The Mediaeval Academy of America Publication 41. Cambridge, MA: Mediaeval Academy.

KIERDORF, WILHELM 1980. *Laudatio funebris: Interpretationen und Untersuchungen zur Entwicklung der römischen Leichenrede.* Beiträge zur Klassischen Philologie 106. Meisenheim a. Glan: Hain.

KIRSCH, JOHANN PETER 1918. *Die römischen Titelkirchen im Altertum.* Studien zur Geschichte und Kultur des Altertums 9, 1/2. Paderborn: Schöningh.

KIRSCH, JOHANN PETER 1924. *Der stadtrömische christliche Festkalender im Altertum: Textkritische Untersuchungen zu den römischen 'Depositiones' und dem Martyrologium Hieronymianum.* Liturgiegeschichtliche Quellen 7/8. Münster: Aschendorff.

KIRSCH, JOHANN PETER 1926. *Die Stationskirchen des Missale Romanum: Mit einer Untersuchung über Ursprung und Entwicklung der liturgischen Stationsfeier.* Ecclesia Orans 19. Freiburg i. Br.: Herder.

KLAUSER, THEODOR 1960. Christlicher Märtyrerkult, heidnischer Heroenkult und spätjüdische Heiligenverehrung: Neue Einsichten und neue Probleme? *10 Jahre Arbeitsgemeinschaft für Forschung des Landes Nordrhein-Westfalen.* Jahresfeier am 11. Mai

1960 in Düsseldorf. Arbeitsgemeinschaft für Forschung des Landes Nordrhein-Westfalen, Geisteswissenschaften 91. Cologne: Westdeutscher Verlag.

KLINGHARDT, MATTHIAS 1988. *Gesetz und Volk Gottes: Das lukanische Verständnis des Gesetzes nach Herkunft, Funktion und seinem Ort in der Geschichte des Urchristentums.* Wissenschaftliche Untersuchungen zum Neuen Testament 2, 32. Tübingen: Mohr.

KLINGHARDT, MATTHIAS 1991. 'auf daß du den Feiertag heiligest': Sabbat und Sonntag im antiken Judentum und frühen Christentum. In JAN ASSMANN (ed.), *Das Fest und das Heilige: Religiöse Kontrapunkte zur Alltagswelt*. Studien zum Verstehen fremder Religionen 1. Gütersloh: Mohn, 206–33.

KOCH, CARL 1953. Bemerkungen zum römischen Quirinuskult. *Zeitschrift für Religions- und Geistesgeschichte* 5, 1–25.

KÖVES-ZULAUF, THOMAS 1990. *Römische Geburtsriten*. Zetemata 87. Munich: Beck.

KOLENDO, JERZY 1981. La répartition des places aux spectacles et la stratification sociale dans l'Empire Romain: A propos des inscriptions sur les gradins des amphithéâtres et théâtres. *Ktema* 6, 301–15.

KRÄMER, HANS JOACHIM 1965. Die Sage von Romulus und Remus in der lateinischen Literatur. In HELMUT FLASHAR AND KONRAD GAISER (eds), *Synusia: Festschrift Wolfgang Schadewaldt*. Pfullingen: Neske, 355–402.

KRUSCH, BRUNO 1938. *Studien zur christlich-mittelalterlichen Chronologie: Die Entstehung unserer Zeitrechnung. I. Victorius. Ersatz der fehlerhaften Ausgabe Mommsens in den M. gG. II. Dionysius Exiguus, der Begründer der christlichen Ära*. Abhandlungen der Preußischen Akademie der Wissenschaften, phil.-hist. Kl. 1937/8. Berlin: Akademie.

KÜBLER, BERNHARD 1897. Calata comitia. *RE* 3 (1), 1330–4.

KUBITSCHEK, WILHELM 1915. *Die Kalenderbücher von Florenz, Rom und Leyden*. Denkschriften der kaiserlichen Akademie der Wissenschaften in Wien, phil.-hist. Kl. 57, 3. Vienna: Hölder.

LAFFI, UMBERTO 1967. Le iscrizioni relative all'introduzione nel 9 a. gC. del nuovo calendario della provincia d'Asia. *Studi Classici e Orientali* 16, 5–98.

LANDSBERGER, BENNO 1915. *Der kultische Kalender der Babylonier und Assyrer*. Leipziger Semitistische Studien 6, 1/2. Repr. Leipzig: Hinrichs.

LANGDON, STEPHEN 1935. *Babylonian Menologies and the Semitic Calendars*. Schweich Lectures on Biblical Archaeology 1933. London: Oxford University Press.

LASKER, ARNOLD A. AND LASKER, DANIEL J. 1984. The Jewish Prayer for Rain in Babylonia. *Journal for the Study of Judaism in the Persian, Hellenistic and Roman Period* 15, 123–44.

LASKER, ARNOLD A. AND LASKER, DANIEL J. 1989. Behold, a Moon Is Born! How the Jewish Calendar Works. *Conservative Judaism* 41, 5–19.

LATTE, KURT 1950. Religiöse Begriffe im frührömischen Recht. In von O. GIGON, W. BUCHWALD AND W. KUNKEL (eds), *Kleine Schriften: zu Religion, Recht, Literatur und Sprache der Griechen und Römer*. [*ZSavSt* 67 (1950), 47–61.] Munich: Beck, 329–40.

LATTE, KURT 1960. *Römische Religionsgeschichte*. Handbuch der Altertumswissenschaft 5, 4. Munich: Beck.

LEBEK, WOLFGANG DIETER 1993. Roms Ritter und Roms Plebs in den Senatsbeschlüssen für Germanicus Caesar und Drusus Caesar. *ZPE* 95, 81–120.

LECLERCQ, HENRI 1925. Calendrier. In FERNAND CABROL (ed.), *Dictionnaire d'Archéologie Chrétienne et de Liturgie* 2, 2. Paris: Letouzey et Ané, 1585–93.

LEFEBVRE, GEORGES 1957. *La Révolution Française*. 2nd edn. Peuples et civilisations 13[b]. Paris: Presses Universitaires.

LEHMANN-HARTLEBEN, KARL 1926. *Die Trajanssäule: Ein römisches Kunstwerk zu Beginn der Spätantike. [1:] Text. [2:] Tafeln.* Berlin: de Gruyter.

LEO, FRIEDRICH 1967 [1913]. *Geschichte der römischen Literatur 1: Die archaische Literatur.* Im Anhang: Die römische Poesie in der sullanischen Zeit. Unv. ND. [Berlin 1913.] Darmstadt: Wissenschaftliche Buchgesellschaft, 1967.

LEWY, HILDEGARD AND LEWY, JULIUS 1943. The Origin of the Week and the Oldest West Asiatic Calendar. *Hebrew Union College Annual* 17, 1–152c.

LIBERO, LORETANA de 1992. *Obstruktion: Politische Praktiken im Senat und in der Volksversammlung der ausgehenden römischen Republik (70–49 v. Chr.).* Hermes Einzelschriften 59. Stuttgart: Steiner.

LIETZMANN, HANS 1927. *Petrus und Paulus in Rom: Liturgische und archäologische Studien.* 2nd revised edition. Arbeiten zur Kirchengeschichte 1. Berlin: de Gruyter.

LIGT, L. de 1993. *Fairs and Markets in the Roman Empire: Economic and Social Aspects of Periodic Trade in a Pre-Industrial Society.* Dutch Monographs on Ancient History and Archaeology 11. Amsterdam: Gieben.

LINDERSKI, JERZY 1986. The Augural Law. *ANRW II* 16 (3), 2146–312.

LINTOTT, ANDREW W. 1965. Trinundinum. *CQ* 15, 281–5.

LINTOTT, ANDREW W. 1968. Nundinae and the Chronology of the Late Roman Republic. *CQ* 18, 189–94.

LUCAS, HANS 1938. Martial's Kalendae nataliciae. *CQ* 32, 5–6.

LUCCHESI, GIOVANNI 1978. Ancora sull'antico calendario italico. *Rivista di Storia della Chiesa in Italia* 32, 140–52.

LUCE, T. J. 1967. A Note on Livy 9, 9, 2: Intercalatae ponae usuram. *Hermes* 95, 383–4.

MCGURK, P. 1986. The Metrical Calendar of Hampson. *Analecta Bollandiana* 104, 79–125.

MACMULLEN, RAMSAY 1991. Hellenizing the Romans (2nd Century BC). *Historia* 40, 419–38.

MAGDELAIN, ANDRÉ 1980. Quando rex comitavit fas. *RHDFE* ser. 4 (58), 5–11.

MALITZ, JÜRGEN 1987. Die Kalenderreform Caesars: Ein Beitrag zur Geschichte seiner Spätzeit. *Ancient Society* 18, 103–31.

MALLARDO, DOMENICO 1944–6. Il calendario marmoreo di Napoli. *Ephemeris Liturgica* 58 (1944), 116–77; 59 (1945), 233–94; 60 (1946), 7–26.

MANCINI, GIOACCHINO 1921. Anzio: Scoperta di un calendario romano, anteriore a Giulio Cesare, e di un brano dei fasti consolari e censori, l'uno e gli altri dipinti sopra intonaco. *NSc 1921*, 73–141.

MARCHETTI, PATRICK 1973. La marche du calendrier romain de 203 à 190 (années varr. 551–564). *AC* 42, 473–96.

MARTIN, JOCHEN 1987. *Spätantike und Völkerwanderung.* Oldenbourg Grundriß der Geschichte 4. Munich: Oldenbourg.

MARTIN, J. 1990. Die Macht der Heiligen. In J. MARTIN (ed.), *Christentum und antike Gesellschaft.* Wege der Forschung 649. Darmstadt, 440–74.

MARTINA, MARIO 1981. Aedes Herculis Musarum. *Dialoghi di Archeologia* NS 3 (1), 49–68.

MATHIEZ, ALBERT 1904. *La théophilanthropie et la culte décadaire: Essai sur l'histoire religieuse de la Révolution, 1796–1801.* Reprint. [Paris: Alcan, 1904.] Geneva: Slatkine-Megariotis.

MAUR, HANSJÖRG AUF DER 1983. *Feiern im Rhythmus der Zeit 1: Herrenfeste in Woche und Jahr.* Gottesdienst der Kirche: Handbuch der Liturgiewissenschaft 5, 1. Regensburg: Pustet.

MAUR, HANSJÖRG AUF DER 1994. Feste und Gedenktage der Heiligen. *Feiern im Rhythmus der Zeit II/1*. Gottesdienst der Kirche: Handbuch der Liturgiewissenschaft 6, 1. Regensburg: Pustet, 65–357.

MAURER, MICHAEL 1991. Feste und Feiern als historischer Forschungsgegenstand. *HZ* 253, 101–30.

MAYR, JOACHIM 1955. Der Computus ecclesiasticus. *Zeitschrift für Katholische Theologie* 77, 301–30.

MEIER, CHRISTIAN 1995 [1982]. *Caesar*. Tr. from the German by David McLintock. London: Harper Collins (*Caesar.* Mit einem Nachwort zur Taschenbuchausgabe). [Berlin: Siedler, 1982.] Munich: dtv.

MEINHOLD, JOHANNES 1905. *Sabbat und Woche im Alten Testament*. Forschungen zur Religion und Literatur des Alten und Neuen Testaments 5. Göttingen: Vandenhoeck & Ruprecht.

MEINHOLD, JOHANNES 1909. *Sabbat und Sonntag*. Wissenschaft und Bildung 45. Leipzig: Quelle & Meyer.

MEINHOLD, JOHANNES 1909a. Die Entstehung des Sabbats. *Zeitschrift für Alttestamentliche Wissenschaft* 29, 81–112.

MEINHOLD, JOHANNES 1930. Zur Sabbathfrage. *Zeitschrift für Alttestamentliche Wissenschaft* 48, 121–38.

MEINZER, MICHAEL 1988. Der französische Revolutionskalender und die 'Neue Zeit'. In REINHART KOSELLECK AND ROLF REICHARDT (eds), *Die Französische Revolution als Bruch des gesellschaftlichen Bewußtseins*. Munich: Oldenbourg, 23–60 (–71: discussion).

MEINZER, MICHAEL 1992. *Der französische Revolutionskalender (1792–1805): Planung, Durchführung und Scheitern einer politischen Zeitrechnung*. Ancien Régime, Aufklärung und Revolution 20. Munich: Oldenbourg.

MELONI, PIERO 1954. Ancora sul Calendario romano nell'anno della bataglia di Pidna (168 a. Cr.). *Latomus* 13, 553–68.

MERCKLIN, LUDWIG 1840–1. *De Junio Gracchano commentatio 1–2*. Dorpat: Laakmann.

METZLER, DIETER 1981. Ökonomische Aspekte des Religionswandels in der Spätantike: Die Enteignung der heidnischen Tempel seit Konstantin. *Hephaistos* 3, 27–40.

MEYER, EDUARD 1922. *Caesars Monarchie und das Principat des Pompejus: Innere Geschichte Roms von 66 bis 44 v. Chr.* Reprint of the 3rd edition. [Stuttgart 1922.] Darmstadt: Wissenschaftliche Buchgesellschaft.

MICHELS, AGNES KIRSOPP 1949. The 'Calendar of Numa' and the Pre-Julian Calendar. *TAPhA* 80, 320–46.

MICHELS, AGNES KIRSOPP 1978 [1967]. *The Calendar of the Roman Republic*. Princeton: Princeton University Press. [Repr. Westport: Greenwood, 1978.]

MILLER, JOHN F. 1992. The Fasti and Hellenistic Didactic: Ovid's Variant Aetiologies. *Arethusa* 25, 11–31.

MILLER, JOHN F. 2002. The Fasti: Style, Structure, and Time. In BARBARA WEIDEN BOYD (ed.), *Brill's Companion to Ovid*. Leiden: Brill, 167–96.

MOMIGLIANO, ARNALDO 1969. Il rex sacrorum e l'origine della repubblica. *Quarto contributo alla storia degli studi classici e del mondo antico*, Storia e letteratura 115. Rome: Storia e letteratura, 395–402.

MOMMSEN, THEODOR 1858. *Die römische Chronologie bis auf Caesar*. Berlin: Weidmann.

MOMMSEN, THEODOR 1882. Das Augustische Festverzeichnis von Cumae. *Gesammelte Schriften 4: Historische Schriften 1*. [Hermes 17 (1882), 631–43.] Berlin: Weidmann, 259–70.

MOMMSEN, THEODOR 1887–8. *Römisches Staatsrecht* 3, 1–2. Handbuch der römischen Alterthümer 3, 1–2. Leipzig: Hirzel.

MONARD, JOSEPH 1999. *Histoire du calendrier gaulois: le calendrier de Coligny*. Burillier. Vannes: Editions Burillier.

MORA, FABIO 1990. *Prosopografia Isiaca 1: Corpus prosopographicum religionis Isiacae. 2: Prosopografia storica e statistica del culto Isiaco*. EPRO 113. Leiden: Brill.

MOSSHAMMER, ALDEN A. 2008. *The Easter Computus and the Origins of the Christian Era*. Oxford: Oxford University Press.

MÜLLER, KARL OTFRIED 1828. *Die Etrusker*. 2 vols. Vier Bücher. Eine von der Königlich Preußischen Akademie der Wissenschaften zu Berlin gekrönte Preisschrift. Breslau: Max.

MÜNZER, FRIEDRICH 1910. Fulvius 91. *RE* 7 (1), 265–7.

MÜNZER, FRIEDRICH 1920. *Römische Adelsparteien und Adelsfamilien*. Stuttgart: Metzler. [2nd unaltered edition 1963].

MULLER, PAUL M. AND STEPHENSON, F. R. 1974. The Accelerations of the Earth and Moon from Early Astronomical Observations. In GARY D. ROSENBERG AND STANLEY KEITH RUNCORN (eds), *Growth Rhythms and the History of the Earth's Rotation*. London: Wiley, 459–533.

MURRAY, OSWYN 1985. Symposium and Genre in the Poetry of Horace. *JRS* 75, 39–50.

NADJO, LÉON 1989. *L'argent et les affaires à Rome des origines au 11e siècle avant J.-C.: Étude d'un vocabulaire technique*. Bibliothèque de l'Information grammaticale 16. Louvain: Peeters.

NADJO, LÉON 1991. Remarques linguistiques sur le calendrier républicain. In RAYMOND CHEVALLIER (ed.), *La révolution française et l'antiquité*. Caesarodunum 25 bis. Tours: Centre de Recherches A. Piganiol, 247–72.

NAIDEN, F. S. 2006. *Ancient Supplication*. Oxford: Oxford University Press

NASSEHI, ARMIN 1993. *Die Zeit der Gesellschaft: Auf dem Weg zu einer soziologischen Theorie der Zeit*. Opladen: Westdeutscher Verlag.

NEWLANDS, CAROLE E. 1995. *Playing with Time: Ovid and the Fasti*. Ithaca, NY: Cornell University Press.

NEWTON, ROBERT R[USSELL] 1972. *Medieval Chronicles and the Rotations of the Earth*. Baltimore: Johns Hopkins Press.

NICOLAI, KARL 1963. Feiertage und Werktage im römischen Leben, besonders in der Zeit der ausgehenden Republik und in der frühen Kaiserzeit. *Saeculum* 14, 194–220.

NICOLET, CLAUDE 1980. Économie, société et institutions au IIe siècle av. J.-C.: De la lex Claudia à l'ager exceptus. *Annales: Économies, sociétes, civilisations* 35, 871–94.

NILSSON, NILS MARTIN PERSON 1920. *Primitive Time-reckoning: A Study in the Origins and First Development of the Art of Counting Time Among the Primitive and Early Culture Peoples*. Acta Societatis Humaniorum Litterarum Ludensis 1. Lund: Gleerup.

NILSSON, NILS MARTIN PERSON 1945. Pagan Divine Service in Late Antiquity. *HThR* 38, 63–70.

NILSSON, NILS MARTIN PERSON 1951. *Cults, Myths, Oracles, and Politics in Ancient Greece*. With two appendices: The Ionian phylae; The phratries. Skrifter utgivna av Svenska

Institutet i Athen 8, 1. Lund: Gleerup. [Repr. Studies in Mediterranean Archaeology 44, Göteborg: Paul Åström, 1986].

NOETHLICHS, KARL LEO 1971. Die gesetzgeberischen Maßnahmen der christlichen Kaiser des 4. Jh. gegen Häretiker, Heiden und Juden. Diss. Cologne.

NOETHLICHS, KARL LEO 1986. Heidenverfolgung. *RAC* 13, 1149–90.

NOLLÉ, JOHANNES MATTHIAS 1982. *Nundinas instituere et habere: Epigraphische Zeugnisse zur Einrichtung und Gestaltung von ländlichen Märkten in Afrika und in der Provinz Asia.* Subsidia epigraphica 9. Hildesheim: Olms.

NORDBERG, HENRIC 1963. *Biometrical Notes: The Information on Ancient Christian Inscriptions from Rome Concerning the Duration of Life and the Dates of Birth and Death.* Acta Instituti Romani Finlandiae 2, 2. Helsinki.

NORTH, JOHN A. 1989. Religion in Republican Rome. In FRANK WALBANK et al. (eds), *Cambridge Ancient History 7, 2: The Rise of Rome to 220 B.C.* 2nd edn. Cambridge: University Press, 573–624.

OBERMANN, JULIAN 1956. Calendaric Elements in the Dead Sea Scrolls. *Journal of Biblical Literature* 75, 285–97.

OGILVIE, ROBERT MAXWELL 1965. *A Commentary on Livy: Book 1–5.* Oxford: Clarendon.

OLMSTED, GARRETT 1992. *The Gaulish Calendar: A Reconstruction from the Bronze Fragments from Coligny* ... Bonn: Habelt.

OPPENHEIM, LEO 1969. Divination and Celestial Observation in the Last Assyrian Empire. *Centaurus* 14, 97–135.

ÖSTENBERG, IDA 2009. *Staging the World: Spoils, Captives, and Representations in the Roman Triumphal Procession.* Oxford: Oxford University Press.

OTTO, WALTER 1905–8. *Priester und Tempel im hellenistischen Ägypten.* 2 vols. Leipzig: Teubner.

OZOUF, MONA 1975. Space and Time in the Festivals of the French Revolution. *Comparative Studies in Society and History* 17, 372–84.

OZOUF, MONA 1989 [1976]. *La fête révolutionnaire: 1789–1799.* [Paris: Folio, 21989.] Paris: Gallimard.

PAIS, ETTORE 1909. Le leggi Pinaria-Furia ed Acilia sulla intercalazione sono esistite? *Studi storici per l'antichità classica* 2, 184–213.

PALLOTTINO, MASSIMO 1988 [1985]. *Etruskologie: Geschichte und Kultur der Etrusker.* Tr. from the Italian by STEPHAN STEINGRÄBER. [*Etruscologia,* 7th new edn. Milan: Hoepli, 1985.] Basel: Birkhäuser, 1988.

PALLOTTINO, MASSIMO 1991 [1984]. *A History of Earliest Italy.* Tr. by MARTIN RYLE AND KATE SOPER. London: Routledge. [*Storia della prima Italia,* Milano: Rusconi, 1984.].

PALMER, ROBERT E. A. 1965. The Censors of 312 B.C. and the State Religion. *Historia* 14, 293–324.

PALMER, ROBERT E. A. 1970. *The Archaic Community of the Romans.* Cambridge: Cambridge University Press.

PAOLI, J. 1950. La signification du sigle NP dans les calendriers romains. *REL* 28, 252–79.

PAOLI, J. 1952. Les définitions varroniennes des jours fastes et néfastes. *RHDFE* ser. 4 (30), 293–327.

PAOLI, J. 1954. La notion de temps faste et celle de temps comitial (Varron, De l. gl., VI, 31 et 32.). *REA* 56, 121–49.

PAPE, MAGRIT 1975. Griechische Kunstwerke aus Kriegsbeute und ihre öffentliche Aufstellung in Rom: Von der Eroberung von Syrakus bis in augusteische Zeit. Diss. Hamburg.

PARKE, HERBERT WILLIAM 1977. *Festivals of the Athenians*. London: Thames & Hudson.

PARKER, RICHARD ANTHONY 1950. *The Calendars of Ancient Egypt*. Studies in Ancient Oriental Civilization 26. Chicago: Chicago University Press.

PERUZZI, EMILIO 1993. La poesia conviviale di Roma arcaica. *Parola di Passato* 272, 332–73.

PFAFF-REYDELLET, MAUD 2009. Ovids Fasti: Der Kaiser tritt in den öffentlichen Kalender ein. In A. BENDLIN AND J. RÜPKE (eds), *Römische Religion im historischen Wandel: Diskursentwicklung von Plautus bis Ovid*. Stuttgart, 157–70.

PFIFFIG, AMBROS JOSEF 1975. *Religio Etrusca*. Graz: Akademische Druck- und Verlagsanstalt.

PHILIP, ALEXANDER 1921. *The Calendar: Its History, Structure and Improvement*. Cambridge: Cambridge University Press.

PICCALUGA, GIULIA 1974. *Terminus: I segni di confine nella religione romana*. Quaderni di SMSR 9. Rome: Ateneo.

PICCALUGA, GIULIA 1977. Irruzione di un passato irreversibile nella realtà cultuale romana. *Studi storico religiosi* 1, 47–62.

PIETILÄ-CASTRÉN, LEENA 1987. *Magnificentia publica: The Victory Monuments of the Roman Generals in the Era of the Punic Wars*. Commentationes Humanarum Litterarum 84. Helsinki: Societas scientiarum Fennica.

PIETRI, CHARLES 1983. Liturgie, Kultur und Gesellschaft: Das Beispiel Roms in der ausgehenden Antike (4.-5. Jahrhundert). *Concilium* 19, 116–24.

PIETRI, CHARLES 1984. Le temps de la semaine à Rome et dans l'Italie chrétienne (Ive–VIe siècle). *Le temps chrétien de la fin de l'antiquité au moyen âge, IIIe–XIIIe siècles*. Paris, 9–12 March 1981. Colloques internationaux du Centre national de la recherche scientifique 604. Paris: CNRS, 63–97.

PIETRI, CHARLES 1984a. Les origines du culte des martyrs (d'après un ouvrage recent). *Rivista di Archeologia Cristiana* 3, 293–319.

PIGHI, GIOVANNI BATTISTA 1965. *De ludis saecularibus populi Romani Quiritium: Libri sex*. Editio altera addendis et corrigendis aucta. [Pubblicazioni dell'Università del s. cuore ser. 5, 25. Milan, 1941.] Amsterdam: Schippers.

POLIAKOFF, DENALI MARIE ST. AMAND 1989. The Acculturation of Jews in the Roman Empire: Evidence from Burial Places. Diss. Boston University.

POMA, GABRIELLA 1984. *Tra legislatori e tiranni: Problemi storici e storiografici sull'età delle XII tavole*. Studi di storia 2. Bologna: Pàtron.

PORTE, DANIELLE 1981. Romulus-Quirinus, prince et dieu, dieu des princes: Étude sur le personnage de Quirinus et sur son évolution, des origines à Auguste. *ANRW II* 17 (1), 300–42.

PORTE, DANIELLE 1989. *Les donneurs de sacré: Le prêtre à Rome*. Paris: Belles Lettres.

PORTE, DANIELLE 1993. Les trois mythologies des 'Fastes'. In GRAF 1993, 142–57.

PRICE, SIMON R. F. 1984. *Rituals and Power: The Roman imperial cult in Asia Minor*. Cambridge: Cambridge University Press.

PRITCHETT, W. K. 1999. Postscript: The Athenian Calendars. *ZPE* 128, 79–93.

QUAEGEBEUR, JAN 1979. Documents égyptiens et rôle économique du clergé en Égypte hellénistique. In EDWARD LIPINSKI (ed.), *State and Temple Economy in the Ancient Near*

East 2. Proceedings of the International conference organized by the Katholieke Universiteit Leuven from the 10th to the 14th of April 1978. Orientalia Lovaniensia Analecta 6. Leuven: Departement Oriëntalistiek, 707–29.

QUASS, FRIEDEMANN 1993. *Die Honoratiorenschicht in den Städten des griechischen Ostens: Untersuchungen zur politischen und sozialen Entwicklung in hellenistischer und römischer Zeit*. Stuttgart: Steiner.

RADKE, MAX GEORG GERHARD 1990. *Fasti Romani: Betrachtungen zur Frühgeschichte des römischen Kalenders*. Orbis antiquus 31. Münster: Aschendorff.

RADKE, MAX GEORG GERHARD 1991. Vergil und Horaz: Ihre Geburtstage: Corrigendum zu Gymnasium 71, 1964, 80 ff. *Gymnasium* 98, 351.

RAEPSAET-CHARLIER, MARIE-THÉRÈSE AND DEMAN, ALBERT 1974. Notes de chronologie romaine. *Historia* 23, 271–96.

RAWSON, ELISABETH 1991. *Roman Culture and Society: Collected Papers*. Oxford: Clarendon.

REA, JOHN R. 1992. A New Bithynian Month List. *ZPE* 90, 289–91.

REEVES, BARBARA M. 2004. The Feriale Duranum, Roman Military Religion, and Dura-Europos: A Reassessment. Diss. NYSU Buffalo.

RICHARDSON, LAWRENCE jr 1977. Hercules Musarum and the Porticus Philippi in Rome. *American Journal of Archaeology* 81, 355–61.

RICHARDSON, LAWRENCE jr 1991. Urban Development in Ancient Rome and the Impact of Empire. In ANTHONY MOLHO, KURT RAAFLAUB AND JULIA EMLEN (eds), *City States in Classical Antiquity and Medieval Italy*. Stuttgart: Steiner, 381–402.

RICHARDSON, LAWRENCE jr 1992. *A New Topographical Dictionary of Ancient Rome*. Baltimore: Johns Hopkins University Press.

RILINGER, ROLF 1976. *Der Einfluß des Wahlleiters bei den römischen Konsulwahlen von 366 bis 50 v. Chr.* Vestigia 24. Munich: Beck.

ROBBINS, ELLEN 1989. Studies in the Prehistory of the Jewish calendar. Diss. New York University.

ROHDE, GEORG 1932. Die Bedeutung der Tempelgründungen im Staatsleben der Römer. In BERNHARD KYTZLER von IRMGARD ROHDE (ed.), *Studien und Interpretationen zur antiken Literatur, Religion und Geschichte*. Berlin: de Gruyter, 189–205.

ROMANO, ANGELA 1990. *Il 'collegium scribarum': Aspetti sociali e giuridici della produzione letteraria tra III e II secolo a. C.* Pubblicazioni del Dipartimento di Diritto Romano e Storia della Scienza Romanistica dell'Università degli Studi di Napoli 'Federico II' 3. Naples: Jovene.

RONCALLI, FRANCESCO 1980. 'Carbasinis voluminibus implicati libri': Osservazioni sul liber linteus di Zagabria. *JdI* 95, 227–64.

RONCALLI, FRANCESCO 1980a. Osservazioni sui libri lintei etruschi. *Pontificia Accademia Romana di Archeologia: Rendiconti* 62 (1978–80), 3–21.

RORDORF, WILLY 1968 [1962]. *Sunday: The History of the Day of Rest and Worship in the Earliest Centuries of the Christian Church*. Tr. from the German by A. A. K. Graham. London: SCM Press. [*Der Sonntag: Geschichte des Ruhe- und Gottesdiensttages im ältesten Christentum*. Abhandlungen zur Theologie des Alten und Neuen Testaments 43. Zürich: Zwingli-Verlag, 1962.]

RORDORF, WILLY 1972. *Sabbat und Sonntag in der alten Kirche*. Traditio Christiana 2. Zürich: Theologischer Verlag.

RORDORF, WILLY 1981. Ursprung und Bedeutung der Sonntagsfeier im frühen Christentum: Der gegenwärtige Stand der Forschung. *Liturgisches Jahrbuch* 31, 145–58.

ROSE, K. F. C. 1971. *The Date and Author of the Satyricon*. Intro. by J. P. SULLIVAN. Mnemosyne suppl. 16. Leiden: Brill.

ROSEN, KLAUS 1985. Die falschen Numabücher: Politik, Religion und Literatur in Rom 181 v. Chr. *Chiron* 15, 65–90.

ROSENBERG, ARTHUR 1914. Rex sacrorum. *RE* 1A, 1, 721–6.

ROTONDI, GIOVANNI 1922 [1912]. *Leges publicae populi romani. Elenco cronologico con una introduzione sull'attivita' legislativa dei comizi romani.* Estratto dalla Enciclopedia giuridica italiana. [Repr. of the Milan edition, 1912, with a postscript by G. Rotondi. Scritti giuridici, I, Milan, 1922.]

RÜPKE, JÖRG 1990. Gerechte Kriege – gerächte Kriege: Die Funktion der Götter in Caesars Darstellung des Helvetierfeldzuges (Gall. 1,12). *AU* 33 (5), 5–13.

RÜPKE, JÖRG 1990a. *Domi militiae: Die religiöse Konstruktion des Krieges in Rom.* Stuttgart: Steiner.

RÜPKE, JÖRG 1991. Dies endotercisi? *ZPE* 86, 212–14.

RÜPKE, JÖRG 1992. You Shall not Kill: Hierarchies of Norms in Ancient Rome. *Numen* 39, 58–79.

RÜPKE, JÖRG 1992a. Wer las Caesars bella als commentarii? *Gymnasium* 99, 201–26.

RÜPKE, JÖRG 1992b. [Review] Gerhard Radke, Fasti Romani: Betrachtungen zur Frühgeschichte des römischen Kalenders ... 1990 ... *Gnomon* 64, 139–44.

RÜPKE, JÖRG 1993. *Römische Religion bei Eduard Norden: Die 'Altrömischen Priesterbücher' im wissenschaftlichen Kontext der dreißiger Jahre.* Religionswissenschaftliche Reihe 7. Marburg: diagonal.

RÜPKE, JÖRG 1993a. Du sollst nicht töten: Hierarchisierung von Normen und ihre rituelle Sanktionierung am Beispiel Roms. In HORST BÜRKLE (ed.), *Grundwerte menschlichen Verhaltens in den Religionen.* Religionswissenschaft 6. Frankfurt a. M.: Lang.

RÜPKE, JÖRG 1993b. Livius, Priesternamen und die annales maximi. *Klio* 74, 155–79.

RÜPKE, JÖRG 1993c. Vexillum caeruleum. *RhM* 136, 374–6.

RÜPKE, JÖRG 1993d. Vergils Laokoon. *Eranos* 91, 126–8.

RÜPKE, JÖRG 1994. Ovids Kalenderkommentar: Zur Gattung der libri fastorum. *Antike und Abendland* 40, 125–36.

RÜPKE, JÖRG 1994a. Wann feierte Ovid die Feralia? Zu Ov. fast. 2,567 f. MH 51, 97–102.

RÜPKE, JÖRG 1994b. Fehler und Fehlinterpretationen in der Datierung des dies natalis des stadtrömischen Mater-Magna-Tempels. *ZPE* 102, 237–40.

RÜPKE, JÖRG 1994c. Die religiöse Konstruktion des Krieges in Rom. In HANS WISSMANN (ed.), *Krieg und Religion.* Würzburg: Königshausen & Neumann, 55–78.

RÜPKE, JÖRG 1995. Fasti: Quellen oder Produkte römischer Geschichtsschreibung. *Klio* 77, 184–202.

RÜPKE, JÖRG 1995a. *Kalender und Öffentlichkeit: Die Geschichte der Repräsentation und religiösen Qualifikation von Zeit in Rom.* Religionsgeschichtliche Versuche und Vorarbeiten 40. Berlin: de Gruyter.

RÜPKE, JÖRG 1995b. Wege zum Töten, Wege zum Ruhm: Krieg in der römischen Republik. In HEINRICH von STIETENCRON (ed.), *Töten im Krieg.* Veröffentlichungen des Instituts für Historische Anthropologie 6. Freiburg: Alber, 211–38.

RÜPKE, JÖRG 1995c. [Review] Michele Renee Salzman, On Roman Time ... 1990 ... *Numen* 42, 210–14.

RÜPKE, JÖRG 1996. Quis vetat et stellas: Les levers des étoiles et la tradition calendaire chez Ovide. In BÉATRICE BAKHOUCHE, ALAIN MOREAU AND JEAN-CLAUDE TURPIN (eds),

Les Astres 1: Les astres et les mythes, la description du ciel. Actes du Colloque International de Montpellier, 23–5 March 1995, Montpellier: Publications de la Recherche, Université Paul Valéry, 293–306.

RÜPKE, JÖRG 1997. Kognitive Einheit ritueller Sequenzen? Zur kommunikativen Funktion kalendarischer Gattungen in Rom. In GERHARD BINDER AND KONRAD EHLICH (ed.), *Stätten und Formen der Kommunikation im Altertum 6: Religiöse Kommunikation – Formen und Praxis vor der Neuzeit*. Bochumer Altertumswissenschaftliches Colloquium 26. Trier: Wissenschaftlicher Verlag Trier, 191–223.

RÜPKE, JÖRG 1998. Les archives des petits collèges: Le cas des vicomagistri. *La mémoire perdue [2]: Recherches sur l'administration romaine*, Rom: Ecole francaise de Rome, 27–44.

RÜPKE, JÖRG 2001. Kulturtransfer als Rekodierung: Zum literaturgeschichtlichen und sozialen Ort der frühen römischen Epik. In RÜPKE (ed.), *Von Menschen und Göttern erzählen: Formkonstanzen und Funktionswandel vormoderner Epik*. Potsdamer altertumswissenschaftliche Beiträge 4. Stuttgart: Steiner, 42–64.

RÜPKE, JÖRG 2004. Acta aut agenda: Relations of script and performance in Roman history of religion. In ALESSANDRO BARCHIESI, JÖRG RÜPKE AND SUSAN STEPHENS (eds), *Rituals in Ink: A Conference on Religion and Literary Production in Ancient Rome, Held at Stanford University in February 2002*. Potsdamer altertumswissenschaftliche Beiträge 10. Stuttgart: Steiner, 23–43.

RÜPKE, JÖRG 2006. Ennius' fasti in Fulvius' Temple: Greek Rationality and Roman Tradition. *Arethusa* 39, 489–512.

RÜPKE, JÖRG 2006a. Tempel, Daten, Rituale: Die vielen Götter als Langzeitgedächtnis der Gesellschaft. In ELKE STEIN-HÖLKESKAMP AND KARL-JOACHIM HÖLKESKAMP (eds), *Erinnerungsorte der Antike: Die römische Welt*. Munich: Beck, 554–69, 768–70.

RÜPKE, JÖRG 2006b. *Zeit und Fest: Eine Kulturgeschichte des Kalenders*. Munich: Beck.

RÜPKE, JÖRG 2007 [2005]. Divination et décisions politiques dans la République romaine. *Cahiers Glotz* 16 (2005), 217–33.

RÜPKE, JÖRG 2007a. *Religion of the Romans*. Tr. and ed. by RICHARD GORDON. Cambridge: Polity Press.

RÜPKE, JÖRG 2007b. The Role of the Roman Calendar for the Formation of the Imperium Romanum and for the Confessional States of the 16th Century. JEAN-PHILIPPE GENET (ed.), *Rome et l"etat modern européen*, Rome: Ecole francaise.

RÜPKE, JÖRG 2008. *Fasti sacerdotum: A Prosopography of Pagan, Jewish, and Christian Religious Officials in the City of Rome, 300 BC to AD 499. Biographies of Christian Officials by* ANNE GLOCK. Tr. by DAVID RICHARDSON. Oxford: Oxford University Press.

RÜPKE, JÖRG 2009. Between Rationalism and Ritualism: On the Origins of Religious Discourse in the Late Roman Republic. *Archiv für Religionsgeschichte* 22, 123–43.

RÜPKE, JÖRG 2009a. Equus October und ludi Capitolini: Zur rituellen Struktur der Oktoberiden und ihren antiken Deutungen. In UELLI DUELL AND CHRISTINA WALDE (eds), *Antike Mythen: Medien, Transformationen und Konstruktionen: FS Fritz Graf zum 65. Geburtstag*. Berlin: de Gruyter, 97–121.

RÜPKE, JÖRG 2009b. Properz: Aitiologische Elegie in Augusteischer Zeit. In ANDREAS BENDLIN AND JÖRG RÜPKE (eds), *Römische Religion im historischen Wandel: Diskursentwicklung von Plautus bis Ovid*. Potsdamer altertumswissenschaftliche Beiträge 17. Stuttgart, 115–42.

RÜPKE, JÖRG AND SPICKERMANN, WOLFGANG (eds) 2009. Religion and Literature. *Archiv für Religionsgeschichte* 11, 121–2.

SABBATUCCI, DARIO 1978. *Il mito, il rito, e la storia*. Rome: Bulzoni.

SABBATUCCI, DARIO 1988. *La religione di Roma antica dal calendario festivo all'ordine cosmico*. La cultura 67. Milan: Saggiatore.

SALZMAN, MICHELE RENEE 1990. *On Roman Time: The Codex-Calendar of 354 and the Rhythms of Urban Life in Late Antiquity*. The Transformation of the Classical Heritage 17. Berkeley: University of California Press.

SALZMAN, MICHELE RENEE 1992. How the West Was Won: The Christianization of the Roman Aristocracy in the West in the Years after Constantine. In CARL DEROUX (ed.), *Studies in Latin Literature and Roman History 6*. Collection Latomus 217. Brussels: Latomus, 451–79.

SALZMAN, MICHELE RENEE 1999. The Christianization of Sacred Time and Sacred Space. In W.V. HARRIS (ed.), *The Transformations of urbs Roma in Late Antiquity. Journal of Roman Archaeology* supple. 33. Portsmouth, RI: JRA, 123–34.

SAMUEL, ALAN E. 1972. *Greek and Roman Chronology: Calendars and Years in Classical Antiquity*. Handbuch der Altertumswissenschaft 1, 7. Munich: Beck.

SBLENDORIO CUGUSI AND MARIA TERESA (eds) 1982. *M. Porci Catonis Orationum reliquiae*. Intro., testo critico e commento filologico. Historica, Politica, Philosophica 12. Turin: Paravia.

SCHANZ, MARTIN AND HOSIUS, CARL 1927. *Geschichte der römischen Literatur bis zum Gesetzgebungswerk des Kaisers Justinian 1: Die römische Literatur in der Zeit der Republik*. 4th revised edition by CARL HOSIUS. ND. [41927.] Handbuch der Altertumswissenschaft 8, 1. Munich: Beck.

SCHEID, JOHN 1975. *Les frères arvales: Recrutement et origine sociale sous les Julio-Claudiens*. Collection de l'École Pratique des Hautes Études, Sciences religieuses 77. Paris: Belles Lettres.

SCHEID, JOHN 1990. *Romulus et ses frères: Le collège des frères arvales, modèle du culte public dans la Rome des empereurs*. Bibliothèque des Écoles françaises d'Athènes et de Rome 275. Rome: École française.

SCHEID, JOHN 1990a. *Le collège des frères arvales: Étude prosopographique du recrutement (69–304)*. Saggi di storia antica 1. Rome: Bretschneider.

SCHEID, JOHN 1990b. Rituel et écriture à Rome. In ANNE-MARIE BLONDEAU AND KRISTOFER SCHIPPER (eds), *Essais sur le rituel 2*. Bibliothèque de l'École des Hautes Études, Sciences religieuses 95. Louvain: Peeters, 1–15.

SCHEID, JOHN 1992. Myth, Cult and Reality in Ovid's Fasti. *Proceedings of the Cambridge Philological Society* 38, 118–31.

SCHEID, JOHN 1993. Die Parentalien für die verstorbenen Caesaren als Modell für den römischen Totenkult. *Klio* 75, 188–201.

SCHEID, JOHN 1998. *Commentarii fratrum Arvalium qui supersunt: Les copies épigraphiques des protocoles annuels de la confrérie arvale (21 av.-304 ap. J.-C.)*. Avec la collaboration de Paola Tassini, Jörg Rüpke. Roma antica 4. Recherches archéologiques à La Magliana. Rome: École française de Rome / Soprintendenza archeologica di Roma.

SCHLAG, URSULA 1968. *Regnum in senatu: Das Wirken römischer Staatsmänner von 200 bis 191 v. Chr.* Kieler Historische Studien 4. Stuttgart: Klett.

SCHNEGG-KÖHLER, BÄRBEL 2002. *Die augusteischen Säkularspiele*. Archiv für Religionsgeschichte 4. Munich: Saur.

SCHOLZ, UDO W. 1984. 'Der Scipio' des Ennius. *Hermes* 112, 183–99.

SCHOLZ, UDO W. 1990. Zur Erforschung des römischen Kalenders. *Orientalia* 59, 255–64.

SCHÖN, GEORG 1909. 'Fasti. II'. *RE* 6 (2), 2023–46.

SCHÜRER, EMIL 1905. Die siebentägige Woche im Gebrauche der christlichen Kirche der ersten Jahrhunderte. *Zeitschrift für die Neutestamentliche Wissenschaft* 6, 1–66.

SCOTT, KENNETH 1931. Greek and Roman Honorific Months. *YCS* 2, 199–278.

SEGUIN, ROGER 1988. Remarques sur les origines des pontifes romains: Pontifex Maximus et Rex Sacrorum. In Porte 1988, 405–18.

SESTON, WILLIAM 1951. Review of T. Robert S. Broughton, *The Magistrates of the Roman Republic*, vol. 1. *REL* 29, 434–6.

SHACKLETON BAILEY, DAVID ROY 1977. *Cicero, Epistulae ad Familiares*. 2 vols. Ed. with comm. Cambridge: Cambridge University Press, 1977.

SHATZMAN, ISRAEL 1972. The Roman General's Authority over Booty. *Historia* 21, 177–205.

SIMONE, CARLO de 1980. Italien. In GÜNTER NEUMANN AND JÜRGEN UNTERMANN (eds), *Die Sprachen im römischen Reich der Kaiserzeit: Kolloquium vom 8. bis 10. April 1974*. Beihefte der BJ 40. Cologne: Rheinland-Verlag, 65–81.

SKUTSCH, OTTO 1985. *The Annals of Q. Ennius*. Oxford: Clarendon.

SMITH, RICHARD J. 1991. *Fortune-tellers and Philosophers: Divination in Traditional Chinese Society*. Boulder: Westview Press.

SOLTAU, WILHELM 1889. *Römische Chronologie*. Freiburg: Mohr.

SOROKIN, PITRIM ALEXANDROWITSCH 1964 [1943]. *Sociocultural Causality, Space, Time: A Study of Referential Principles of Sociology and Social Science*. [Repr. Durham, NC: Duke University Press, 1943.] New York: Russel & Russel, 1964.

SOROKIN, PITRIM ALEXANDROWITSCH AND MERTON, ROBERT K. 1937. Social Time: A Methodological and Functional Analysis. *American Journal of Sociology* 42, 615–29.

SPIER, ARTHUR 1952. *The Comprehensive Hebrew Calendar: Its Structure, History and One Hundred Years of Corresponding Dates, 5600–5760/1900–2000*. New York: Behrman, 1952.

STAVELEY, E. STUART 1959. The Political Aims of Appius Claudius Caecus. *Historia* 8, 410–33.

STEIN, PAUL 1930. Die Senatssitzungen der Ciceronischen Zeit (68–43). Diss. Münster.

STERN, SACHA 2001. *Calendar and Community: A History of the Jewish Calendar, Second Century BCE to Tenth Century CE*. Oxford: Oxford University Press.

STERN, SACHA 2003. *Time and Process in Ancient Judaism*. Oxford: Littman Library of Jewish Civilization.

STROBEL, AUGUST 1977. *Ursprung und Geschichte des frühchristlichen Osterkalenders*. Texte und Untersuchungen zur Geschichte der altchristlichen Literatur 121. Münster: Aschendorff.

STROBEL, AUGUST 1984. *Texte zur Geschichte des frühchristlichen Osterkalenders*. Liturgiewissenschaftliche Quellen und Forschungen 64. Münster: Aschendorff.

STUIBER, ALFRED 1960. Heidnische und christliche Gedächtniskalender. *JbAC* 3, 24–33.

SUERBAUM, WERNER 1968. *Untersuchungen zur Selbstdarstellung älterer römischer Dichter: Livius Andronicus, Naevius, Ennius. Spudasmata 19*. Hildesheim: Olms.

SUERBAUM, WERNER 1980. Merkwürdige Geburtstage. *Chiron* 10, 327–55.

SYDENHAM, M. J. 1974. *The First French Republic, 1792–1804*. London: Batsford.

SYME, RONALD 1982. *Arval Brethren*. Oxford: Clarendon.

SZEMLER, GEORGE J. 1978. Pontifex. *RE* suppl. 15, 331–96.

TALBERT, RICHARD J. A. 1984. *The Senate of Imperial Rome.* Princeton: Prinecton University Press.

TALLEY, THOMAS JULIAN 1981. Historische Elemente zur Formulierung einer christlichen 'Festkunde'. *Concilium* 17, 110–21.

TALLEY, THOMAS JULIAN 1982. Liturgische Zeit in der Alten Kirche: Der Forschungsstand. *Liturgisches Jahrbuch* 32, 25–45.

TAMM, BIRGITTA 1961. Le temple des Muses à Rome. *Opuscula Romana.* Skrifter utgivna av Svenska Institutet i Rom, 4, 21, 157–67.

TAYLOR, LILY ROSS AND HOLLAND, LOUISE ADAMS 1952. Janus and the Fasti. *CP* 47, 137–42.

TELLEGEN-COUPERUS, OLGA 2006 Pontiff, praetor, and iurisdictio in the Roman republic. *Tijdschrift voor Rechtsgeschiedenis* 74, 31–44.

THOMMEN, LUKAS 1989. *Das Volkstribunat der späten römischen Republik.* Historia Einzelschriften 59. Stuttgart: Steiner.

TORELLI, MARIO 1984. *Lavinio e Roma: Riti iniziatici e matrimonio tra archeologia e storia.* Rome: Quasar.

TREBILCO, PAUL R. 1991. *Jewish Communities in Asia Minor.* Society for New Testament Studies. Monograph Series 69. Cambridge: Cambridge University Press.

TROMP, SEBATIAN PETRUS CORNELIS 1921. De Romanorum piaculis. Diss. Leiden.

TURNER, VICTOR W. 1967. *The Forest of Symbols: Aspects of Ndembu Ritual.* Ithaca, NY: Cornell University Press.

TURNER, VICTOR W 1982. Introduction. In TURNER (ed.), *Celebration: Studies in Festivity and Ritual.* Washington, DC: Smithsonian Institution Press.

TURTON, DAVID AND RUGGLES, CLIVE 1978. Agreeing to Disagree: The Measurement of Duration in a Southwestern Ethiopian Community. *Current Anthropology* 19, 585–600.

ULF, CHRISTOPH 1982. *Das römische Lupercalienfest: Ein Modellfall für Methodenprobleme in der Altertumswissenschaft.* Impulse der Forschung 38. Darmstadt: Wissenschaftliche Buchgesellschaft.

UNGER, GEORG FRIEDRICH 1895. Nundinalfragen. *Jahrbücher für classische Philologie* 41, 495–520.

UNGERN-STERNBERG, JÜRGEN von 1993. Romulus-Bilder: Die Begründung der Republik im Mythos. In GRAF 1993, 88–108.

USENER, HERMANN 1913 [1875]. *Italische Mythen. RhM NF 30,* 182–229. [Repr. *Kleine Schriften 4: Arbeiten zur Religionsgeschichte.* Leipzig: Teubner, 1913, 93–143.]

VAN HAEPEREN, FRANÇOISE 2002. *Le collège pontifical (3ème s.a. C.–4ème s.p. C.): contribution à l'étude de la religion publique romaine.* Brussels/Rome: Institut historique Belge de Rome.

VANSINA, JAN 1985. *Oral Tradition as History.* London/Nairobi: Currey/Heinemann.

VEYNE, PAUL 1979. *The Hellenization of Rome and the Question of Acculturations. Diogenes* June (106), 1–27.

VEYNE, PAUL 1990 [1976]. *Bread and Circuses: Historical Sociology and Political Pluralism.* Abridged with an introduction by OSWYN MURRAY. Tr. from the French by BRIAN PEARCE. London: Allen Lane. [*Le Pain et le cirque: Sociologie historique d'un pluralisme politique,* Paris: Éditions du Seuil, 1976.]

VOGEL, CYRILLE 1981. *Medieval Liturgy: An Introduction to the Sources*. Rev. and tr. by WILLIAM G. STOREY AND NIELS KROGH RASMUSSEN with the assistance of JOHN K. BROOKS-LEONARD [*Introduction aux sources de l'histoire du culte chrétien au moyen âge*, Spoleto: Centro italiano di studi sull'alto medioevo, 1981]. Washington, DC: Pastoral Press.

THOMAS VOGTHERR 2001. *Zeitrechnung – von den Sumerern bis zur Swatch*. Munich: Beck.

VON DEN BRINCKEN, ANNA-DOROTHEE 1979. Beobachtungen zum Aufkommen der retrospektiven Inkarnationsära. *Archiv für Diplomatik* 25, 1–20.

WACKERNAGEL, JAKOB 1924. Dies ater. *ARW* 22, 215–16. [Reprint *Kleine Schriften 2*. Göttingen: Vandenhoeck & Ruprecht, o. gJ. (1955, [2]1969), 1298–9.]

WALDE, ALOIS AND HOFMANN, J. B. 1938–54. *Lateinisches etymologisches Wörterbuch*. 3rd revised edition by J. B. HOFMANN. 2 vols. Heidelberg: Winter.

WALLACE-HADRILL, ANDREW 1988. Time for Augustus: Ovid, Augustus and the Fasti. In MICHAEL WHITBY, PHILIP HARDIE AND MARY WHITBY (eds), *Homo Viator: Classical Essays for John Bramble*. Bristol: Classical Press, 221–30.

WALLACE-HADRILL, ANDREW 1990. *Patronage in Ancient Society*. Leicester-Nottingham Studies in Ancient Society 1. London: Routledge.

WARRIOR, VALERIE M. 1991. Notes on Intercalation. *Latomus* 50, 80–7.

WARRIOR, VALERIE M. 1992. Intercalation and the Action of M. Acilius Glabrio (cos. 191 BC). In CARL DEROUX (ed.), *Studies in Latin Literature and Roman History* 6. Collection Latomus 217. Brussels: Latomus, 119–44.

WEBER, GREGOR 1992. Poesie und Poeten an den Höfen vorhellenistischer Monarchen: *synēsan dè ára kaì tóte basileúsi poiētaí* (Paus. 1,2,3). *Klio* 74, 25–77.

WEBER, GREGOR 1993. *Dichtung und höfische Gesellschaft: Die Rezeption von Zeitgeschichte am Hof der ersten drei Ptolemäer*. Hermes Einzelschriften 62. Stuttgart: Steiner.

WEBSTER, HUTTON 1916. *Rest Days: A Study in Early Law and Morality*. New York: Macmillan.

WEINSTOCK, STEFAN 1971. *Divus Julius*. Oxford: Clarendon.

WHATMOUGH, JOSHUA 1931. The Calendar in Ancient Italy outside Rome. *HSCPh* 42, 157–79.

WIEACKER, FRANZ 1988. *Römische Rechtsgeschichte: Quellenkunde, Rechtsbildung, Jurisprudenz und Rechtsliteratur 1,1: Einleitung, Quellenkunde, Frühzeit und Republik*. Handbuch der Altertumswissenschaft 10, 3, 1, 1. Munich: Beck.

WILLIS, GEOFFREY GRIMSHAW 1968. *Further Essays in Early Roman Liturgy*. Alcuin Club Collections 50. London: SPCK.

WINIARCZYK, M. 1994. *Bibliographie zum antiken Atheismus, 17. Jahrhundert-1990*. Bonn: Habelt.

WISEMAN, TIMOTHY PETER 1995. *Remus: A Roman Myth*. Cambridge: Cambridge University Press, 1995.

WISSOWA, GEORG 1902. *Religion und Kultus der Römer*. 1st edition. Handbuch der Altertumswissenschaft 5, 4. Munich: Beck.

WISSOWA, GEORG 1909. Fasti. I. *RE* 6 (2), 2015–23.

WISSOWA, GEORG 1912. *Religion und Kultus der Römer*. 2nd edition. Handbuch der Altertumswissenschaft 5, 4. Munich: Beck.

WISSOWA, GEORG 1923. Vestalinnenfrevel. *ARW* 22, 201–14.

WISSOWA, GEORG 1923a. Neue Bruchstücke des römischen Festkalenders. *Hermes* 53, 369–92.

WOLF, JOSEPH GEORG 1980. Comitia quae pro collegio pontificum habentur. In KLAUS LUIG AND DETLEF LIEBS (eds), *Das Profil des Juristen in der europäischen Tradition: Symposion aus Anlaß des 70. Geburtstages von Franz Wieacker*. Ebelsbach: Gremer, 1–24.

WOLF, JOSEPH GEORG 1980a. *Die literarische Überlieferung der Publikation der Fasten und Legisaktionen durch Gnaeus Flavius*. Nachrichten der Göttinger Akademie der Wissenschaften: I. Phil.-hist. Kl. 1980, 2. Göttingen: Vandenhoeck & Ruprecht, 9–29.

WOODHEAD, A. GEOFFREY 1989. The Calendar of the Year 304/3 B.C. in Athens. *Hesperia* 58, 297–301.

WORP, KLAAS A. 1991. Remarks on Weekdays in Late Antiquity Occurring in Documentary Sources. *Tyche* 6, 221–30.

ZANKER, PAUL 1968. *Forum Augustum: Das Bildprogramm*. Monumenta artis antiquae 2. Tübingen: Wasmuth.

ZANKER, PAUL 1988. *Pompeji: Stadtbilder als Spiegel von Gesellschaft und Herrschaftsform*. Trierer Winckelmannsprogramme 9. Mainz: Zabern.

ZANKER, PAUL 1990 [1987]. *The Power of Images in the Age of Augustus*. Tr. from the German by Alan Shapiro. Ann Arbor: University of Michigan Press. [*Augustus und die Macht der Bilder*. Munich: Beck, 1987.]

ZECCHINI, GIUSEPPE 2001. *Cesare e il mos maiorum*. Historia Einzelschriften 151. Stuttgart: Steiner.

ZELZER, MICHAELA 1978. Zum Osterfestbrief des heiligen Ambrosius und zur römischen Osterfestberechnung des 4. Jahrhunderts. *Wiener Studien NF* 12, 187–204.

ZERUBAVEL, EVIATAR 1977. The French Republican Calendar: A Case Study in the Sociology of Time. *American Sociological Review* 42, 868–77.

ZIOLKOWSKI, ADAM 1992. *The Temples of Mid-Republican Rome and their Historical and Topographical Context*. Saggi di Storia antica 4. Rome: Bretschneider.

ZION WACHOLDER, BEN AND WEISBERG, DAVID B. 1971. Visibility of the New Moon in Cuneiform and Rabbinic Sources. *Hebrew Union College Annual* 42, 227–42.

ZORZETTI, NEVIO 1990. The Carmina Conviviala. In OSWYN MURRAY (ed.), *Sympotica: A Symposium on the Symposion*. Oxford: Clarendon, 289–307.

ZWIERLEIN, OTTO 1982. Der Ruhm der Dichtung bei Ennius und seinen Nachfolgern. *Hermes* 110, 85–102.

Sources Index

The Roman Calendar from Numa to Constantine: Time, History and the Fasti, First Edition. Jörg Rüpke.

Index compiled by Mihaela Holben

General Index

The Roman Calendar from Numa to Constantine: Time, History and the Fasti, First Edition. Jörg Rüpke.

Index compiled by Mihaela Holben

Printed and bound by CPI Group (UK) Ltd, Croydon, CR0 4YY
16/09/2026
14963461-0008